Interactive Visual Data Analysis

AK Peters Visualization Series

This series aims to capture new developments and summarize what is known over the whole spectrum of visualization by publishing a broad range of textbooks, reference works, and handbooks. It will contain books from all subfields of visualization, including visual analytics, information visualization, and scientific visualization. The scope will largely follow the calls of the major conferences such as VIS: techniques, algorithms, theoretical foundations and models, quantitative and qualitative evaluation, design studies, and applications.

Series Editor: Tamara Munzner

University of British Columbia, Vancouver, Canada

Visualization Analysis and Design
Tamara Munzner

Information Theory Tools for Visualization
Min Chen, Miquel Feixas, Ivan Viola, Anton Bardera, Han-Wei Shen, Mateu Sbert

Data-Driven Storytelling
Nathalie Henry Riche, Christophe Hurter, Nicholas Diakopoulos, Sheelagh Carpendale

Interactive Visual Data Analysis
Christian Tominski, Heidrun Schumann

For more information about this series please visit:
https://www.crcpress.com/AK-Peters-Visualization-Series/book-series/CRCVIS

Interactive Visual
Data Analysis

Christian Tominski
Heidrun Schumann

CRC Press
Taylor & Francis Group
Boca Raton London New York

CRC Press is an imprint of the
Taylor & Francis Group, an **informa** business

AN A K PETERS BOOK

CRC Press
Taylor & Francis Group
6000 Broken Sound Parkway NW, Suite 300
Boca Raton, FL 33487-2742

© 2020 by Taylor & Francis Group, LLC
CRC Press is an imprint of Taylor & Francis Group, an Informa business

No claim to original U.S. Government works

Printed on acid-free paper

International Standard Book Number-13: 978-1-4987-5398-2 (Hardback)
978-0-3678-9875-5 (Paperback)

Visit the Taylor & Francis Web site at
http://www.taylorandfrancis.com

and the CRC Press Web site at
http://www.crcpress.com

*To the Rostock
visualization group.*

Contents

Foreword

Big Data has attracted much attention. Many examples are given in the news how new data-driven applications and algorithms provide new insights, enable more effective and efficient operation, and support experts in a wide variety of application domains. This suggests that getting value out of data is easy, but practitioners in the field know better. Obtaining novel, actionable insights from large and complex data is hard work. In many cases, the human in the loop is indispensable to find unexpected relations, to judge if there is true value in the data and to check if results are valid, as a human is needed to bring in both deep domain expertise and common sense. How to bring the human in the loop?

One route is to use *visualization*: translate data into images and take advantage of the incredible strength of the human visual system, enabling the viewer to detect patterns quickly and leading to deep insights. However, visualization has its limits and for instance larger datasets quickly lead to cluttered images, like the hairballs that result when networks with more than a hundred nodes are visualized. An important next level is to bring in *interaction*: enable people to select the data they are interested in. When carefully designed, such interactive visualization systems enable exploration of much larger datasets. But, how to find relevant relations in collections of millions of items, with hundreds of attributes? Statistics, machine learning, and data mining provide many *analysis* methods geared towards finding patterns and correlations automatically, but such analysis methods inevitably also lead to reduction and possibly the loss of vital information. The credo of *Interactive Visual Data Analysis*, also known as visual analytics, is therefore to use a combination of visualization, interaction, and automated analysis methods to explore huge amounts of data and to obtain solid insights.

Problem solved? Unfortunately not. The design of effective systems for visual data analysis is far from trivial, and requires a wide variety of know-how, skills, and experience. Besides pure technical skills, many other aspects are relevant. For each element of such a system a wide variety of alternative solutions are available, selecting the most suitable requires a thorough understanding of their strengths and limitations. When standard solutions do not suffice, creativity is needed to find new methods or ingenious combinations of existing ones. And, above all, development of such systems is a design process, having a deep understanding of this process and dedication to the needs and wants of the prospective users is essential.

I teach students to develop interactive systems for visual data analysis. And, I struggle with this. Should I guide them through the complete zoo of all different visualization, interaction, and analysis techniques, for all kinds of different data? That would lead to a very lengthy and not really exciting course. What I really missed is a compact textbook that provides an overview of interactive visual data analysis: a book that guides through the design process, discusses the supporting disciplines and how they fit together, and gives examples of more advanced topics.

When Christian Tominski and Heidrun Schumann wrote me that they had prepared such a book, I was very excited, and even more when I read it. This book really fills a gap. There are of course many texts on visualization, interaction, and data analysis, all with their own angle, but an integrated overview is rare, whereas I think that the real magic comes from the combination. The authors have an extensive experience in doing research in visualization and developing real-world solutions for real-world problems, and that shows. Many examples come from their own work and give us an opportunity to look over the shoulders of experts. They have a broad experience with many different topics and areas, and have attacked many complex cases, such as dynamic geo-spatial data. I therefore expect that many readers can take direct advantage of the ideas, solutions, and examples given.

Besides their experience in research and development, the authors also have a solid track record in communicating their insight in the field: their book "Visualization of Time-Oriented Data", together with Wolfgang Aigner and Silvia Miksch, is a classic. In the current book, they again show their mastery in handling a complex topic, and I am very impressed by their thoughtful treatment and analysis, their clear and compact writing style, and all clear and useful illustrations.

In short, I highly recommend this book. It provides clear guidance, a thorough overview, and many ideas to students, teachers and everybody who aims to develop effective systems for interactive visual data analysis. I hope this book will inspire many to develop effective solutions for interactive visual data analysis that enable experts to extract great value and deep insights from huge, complex datasets.

Jarke J. van Wijk

Department of Mathematics and Computer Science
Eindhoven University of Technology, The Netherlands
October 2019

Preface

In the year 2000, Heidrun Schumann co-authored the first German textbook on visualization. Christian Tominski was among the first students to use that textbook during their studies. Now, about twenty years later, we, Heidrun and Christian, are happy to present this new book on interactive visual data analysis.

About this Book   The idea for this book was born at IEEE VIS 2014 in Paris during a discussion of Heidrun and Tamara Munzner, who had just published her own book *Visualization Analysis and Design*. The goal was to write a book that provides a comprehensive overview of the principles of interactive visual data analysis. It soon became clear to us that this was not an easy endeavor. Before us were five years of systematization, categorization, harmonization, and over and over again in-depth discussions on how the content could be best prepared, organized, and presented to our readers.

The result is a book with five core chapters. What differentiates our book most from others is its visual analytics focus, that is, the synthesis of visuals, interactivity, and analytics. The book introduces criteria for designing interactive visual data analysis solutions, discusses factors influencing the design, and examines the involved processes. The reader is made familiar with the basics of visual encoding and gets to know numerous visualization techniques for multivariate data, temporal data, geo-spatial data, and graph data. A dedicated chapter introduces general concepts for interacting with visualizations and illustrates how modern interaction technology can facilitate the visual data analysis in many ways. Addressing today's large and complex data, the book covers relevant automatic analytical computations to support the visual data analysis. The book also sheds light on advanced concepts for visualization in multi-display environments, user guidance during the data analysis, and progressive visual data analysis. In her review of this book, Tamara Munzner commented: "I was consistently impressed by how broad the set of things you've done is."

The entire book is richly illustrated with many examples and use cases. Here, we have chosen to rely primarily on our own visual analytics research. This allowed us to create many new schematic depictions and expressive visualizations to illustrate the discussed topics. A great advantage of creating new illustrations is that we could make them available free of charge under Creative Commons Attribution 4.0 International License (CC BY 4.0).

The multitude of topics described in the book can be of interest to a wide range of readers. Students can use the book to learn about interactive visual data analysis. They will benefit most from the structured top-down view offered by the book. Visualization experts can use the book as a reference and for teaching. The advanced concepts discussed at the end of the book may serve as inspiration for new research. Last but not least, domain experts will find the many examples and use cases interesting. Given the broad scope of the book, practitioners from many domains may gain from it.

Acknowledgments   This book might have been written by two people, but it represents the work of many people. Because this book would not exist without our own previous research, special thanks go to all of our former and current colleagues from the Rostock visualization group. Moreover, we would like to say many thanks to our friends and partners with whom we collaborated and co-authored papers that had a substantial impact on this book.

We are also very grateful for support in creating images and transferring rights to use images. Many thanks go to (in alphabetical order) Marco Angelini, Nicolas Belmonte, Thomas Butkiewicz, Steve Dübel, Christian Eichner, Steffen Hadlak, Helwig Hauser, Alexander Lex, Martin Luboschik, Thomas Nocke, Axel Radloff-Delosea, Martin Röhlig, and Sylvia Saalfeld.

Feedback from reviewers greatly helped to improve the book. Many thanks go to Wolfgang Aigner for his review of the chapter on interaction. Martin Luboschik provided valuable comments on the visualization chapter. Hans-Jörg Schulz not only gave us useful feedback on the chapters on fundamental aspects and automatic analysis support as well as on the summary and the preface, he also took part in many discussions and contributed in the early phase of the book project.

Tamara Munzner deserves special thanks for her in-depth review of the entire almost finished book. Her comments helped us a lot in finalizing and polishing the book. Sunil Nair, the editorial director at Taylor & Francis, and his editorial assistants Kirsten Barr and Shikha Garg supported us on the publisher's side.

Finally, we would like to thank our families for providing the warm and heartfelt atmosphere that is necessary to successfully complete a book project. Thank you for your love and support!

Christian Tominski
Heidrun Schumann

Institute for Visual & Analytic Computing
University of Rostock, Germany
October 2019

Authors

Christian Tominski is a researcher and lecturer at the Institute for Visual & Analytic Computing at the University of Rostock, Germany. He received doctoral (Dr.-Ing.) and post-doctoral (Dr.-Ing. habil.) degrees in 2006 and 2015, respectively. His main research interests are in visualization of and interaction with data. He is particularly interested in effective and efficient techniques for interactively exploring and editing complex data. Christian has published numerous papers on new visualization and interaction techniques for multivariate data, temporal data, geo-spatial data, and graphs. He co-authored two books on the visualization of time-oriented data in 2011 and on interaction for visualization in 2015. Christian has developed several visualization systems and tools, including the LandVis system for spatio-temporal health data, the VisAxes tool for time-oriented data, and the CGV system for coordinated graph visualization.

Heidrun Schumann is a professor at the University of Rostock, Germany, where she is heading the Chair of Computer Graphics at the Institute for Visual & Analytic Computing. She received doctoral degree (Dr.-Ing.) in 1981 and post-doctoral degree (Dr.-Ing. habil.) in 1989. Her research and teaching activities cover a variety of topics related to computer graphics, including information visualization, visual analytics, and rendering. She is interested in visualizing complex data in space and time, combining visualization and terrain rendering, and facilitating visual data analysis with progressive methods. A key focus of Heidrun's work is to intertwine visual, analytic, and interactive methods for making sense of data. Heidrun published more than 200 articles in top venues and journals. She co-authored the first German textbook on data visualization in 2000 and a textbook specifically on the visualization of time-oriented data in 2011. In 2014, Heidrun was elected as a Fellow of the Eurographics Association.

Introduction

CONTENTS

D ATA have become a most valuable good. Doctors rely on rich databases about diagnoses and medications to give patients the best possible treatments. Enterprises generate profits based on data about needs and preferences of potential customers. Scientists make new discoveries and contribute to a vast body of scholarly data.

Data are everywhere in the information age. Data are collected by huge numbers of devices equipped with various sensors. A smartphone with a dozen and more different sensors is not uncommon. Data are also generated computationally. Sophisticated models are constructed and simulated in an attempt to estimate how our climate may look like in a few centuries. And we as humans are sources of data as well. Social networks and messaging services record our interests and daily activities.

Now with so much data available, the question is how can we make sense of them? Well, the data have to be explored and analyzed in order to derive valuable information. To this end, a channel has to be established for the data to enter into the human mind where insight can be generated. The classic way of ingesting data is to decipher alphanumerically encoded transcripts, or simply texts. Yet, reading piles of documents is too time consuming. Therefore, data are often aggregated in reports, which may contain structured tabular information. This already helps in extracting the key messages. But complex relationships within the data may still be too difficult to identify.

This is where interactive visual data analysis enters the stage. While text and reports are serial media, where one piece of data has to be processed after the other, visual methods aim for the human visual system at its full bandwidth. Humans are amazingly fast in extracting information from graphical depictions. Graphical means are not only beneficial for communicating information, they also serve as scaffolds for human sensemaking. Mental models can be established more easily with the help of visual abstractions and visually acquired information can be remembered better than textual descriptions.

We all know the idiom: "A picture is worth a thousand words." But this is not quite right. More correct would be "A picture *can be* worth *hundreds of thousands of* words." This slightly provocative statement hints at two important aspects. First, *can be* suggests that there are not only good visual representations of data but also exemplars that are not so helpful. Second, the increase in the number of words is to indicate that, in the information age, we are facing *big data*.

This book is about concepts and methods for the interactive visual analysis of large and complex data by jointly exploiting the power of humans and computers. In this book, you will learn that solutions for interactive visual data analysis are not created in passing. Careful design is necessary before expressive visual representation can be shown on a computer display. Useful interaction is essential to enable users to engage in a dialog with the data and the information contained therein. Especially in the light of big data, we need support from analytic computations to help us extract interesting features from the data.

1.1 BASIC CONSIDERATIONS

Before we go into any details about interactive visual data analysis, let us briefly look at some fundamental terms, ideas, and concepts.

1.1.1 Visualization, Interaction, and Computation

Visualization is a computational process that generates visual representations of data. A first definition has been established by visualization pioneers in 1987. Their definition reads as follows [MDB87]:

> "Visualization is a method of computing. It transforms the symbolic into the geometric, enabling researchers to observe their simulations and computations. Visualization offers a method to see the unseen. It enriches the process of scientific discovery and fosters profound and unexpected insights."
>
> McCormick et al., 1987

This definition describes a transformation that involves several entities and steps. First, there are the data into which we seek insight. Second, there is the computer that transforms the data into visual representations. Finally, there is the human who is making sense of the visual representation.

Visual representations are fundamental for visually driven data-intensive work, but they alone can hardly satisfy all the analytic needs we are facing in the information age. We need support from interaction mechanisms and computational analysis methods.

Already in 1981, Bertin recognized the need for interactively adjustable visual representations [Ber81]:

"A graphic is not 'drawn' once and for all; it is 'constructed' and reconstructed until it reveals all the relationships constituted by the interplay of the data. The best graphic operations are those carried out by the decision-maker himself."

Bertin, 1981

Interaction adds the necessary flexibility to visualization. It allows us to actively take part in the visual data analysis. We may want to focus on different features of the data, look at the data from different perspectives, or adjust visual representations so as to crystallize the desired insights.

While interaction incorporates human competences into the sense-making process, automatic computational methods utilize the power of the machine. Large and complex data usually cannot be visualized in their entirety. Automatic computational analyses crunch the data in search for characteristic features or meaningful abstractions that are easier to digest than the raw data.

The important interplay of visualization, interaction and computational analysis is summarized in the *Visual Analytics Mantra* by Keim and colleagues [Kei+06]:

"Analyse First –
Show the Important –
Zoom, Filter and Analyse Further –
Details on Demand"

Keim et al., 2006

According to this mantra, visual analysis starts with an automatic analytic phase. The important features extracted in this phase are then visualized. Via interaction, the visual representation is adjusted, the data are filtered, and further analytic computations are triggered. Details are readily available upon request.

This tight interleaving of computational and human efforts is the key benefit of interactive visual data analysis as a knowledge-generation approach. The computer can process large amounts of data quickly and accurately. The human has enormous pattern-detection abilities and is proficient in creative thinking and flexible decision-making.

A direct consequence of the interplay of data, humans, and computers is that knowledge from different fields has to be brought together for a successful data analysis. Relevant topics include visual design, computer graphics, human-computer interaction, user interfaces, psychology, data science, and algorithms, to name only a few. The need to get diverse methods work in concert makes the development of practical solutions a non-trivial endeavor.

1.1.2 Five Ws of Interactive Visual Data Analysis

In order to come up with helpful data analysis tools, their context of use needs to be taken into account in the first place. One way to describe the context is to follow a variation of the *Five Ws*: What, why, who, where, and when.

What data are to be analyzed? There are many different types of data, such as player statistics, census data, movement trajectories, and biological networks. Each type of data comes with its own individual characteristics, including data scale, dimensionality, and heterogeneity.

Why are the data analyzed? The objective is to help people accomplish their goals, for example, finding governing factors in a gene regulatory network. Goals typically involve a number of analytic tasks, such as identifying data values or setting patterns in relation.

Who will analyze the data? A doctor who studies data in day-to-day clinical routine needs different analysis tools than a strategic investor who is exploring streams of news data in search for new market opportunities. Individual abilities and preferences play a role as well.

Where will the data be analyzed? The regular workplace is certainly the classic desktop setup with a display, mouse, and keyboard. Yet, there are also large display walls and interactive surfaces that offer new opportunities for interactive visual data analysis.

When will the data be analyzed? As with any tool, visualization, interaction, and computation are means that must be at hand at the right time. A data analysis may follow domain-specific workflows where each step is associated with its own individual requirements.

These Five Ws suggest that there are many factors influencing the development of data analysis tools, including data types, analytic tasks, user groups, display environments, domain conventions, and so forth. Factors related to the What and the Why are crucial for the practical applicability of data analysis tools. Certainly, any visually driven and interactively controlled tool has to consider human factors, the Who, with regard to perceptual, cognitive, and physical abilities, expertise, background, and preferences. The Where and When aspects become increasingly relevant when the data analysis runs on multiple heterogeneous displays, supports collaborative sessions, or follows domain-specific workflows.

In the light of the Five Ws it is clear that an interactive visual data analysis solution, in order to be successful, has to be tailored for a specific purpose and setting. Given the wealth of analytic questions we are facing in the information age, a large variety of concepts and techniques is needed. Next, we look at a few introductory examples.

1.2   INTRODUCTORY EXAMPLES

So far, we have sketched the basic idea of interactive visual data analysis on a rather abstract level. In the following, a series of examples will demonstrate the communicative power of visual analysis approaches, on the one hand, and the involved design decisions and challenges, on the other hand.

The examples will take us from basic visual representations to advanced analysis scenarios. On the way, we will increase the degree of sophistication of the examples by enhancing the visual mapping, integrating interaction mechanisms and automatic computations, combining multiple views, incorporating user guidance, and considering multi-display environments.

1.2.1   Starting Simple

The data we will analyze are graphs (the What). Graphs are a general model for describing entities and relations among them. They are universally useful in many different domains. Biologists model natural phenomena via gene regulation networks, climate researchers make use of climate networks to simulate weather on earth, and crime investigators sketch connections between suspects to solve complicated cases. Typical examples from our daily lives are computer networks and social networks.

A graph generally consists of nodes, edges, and attributes. Nodes represent entities, whereas edges represent relationships between the entities. Nodes as well as edges can have attributes that store additional information.

Before we start with visual examples, let us first take a look at the raw data to be visualized. Listing 1.1 shows our graph stored in the JSON format. Lines 2–11 contain three nodes, each associated with an `id` and a `label`. Lines 14–22 define two edges. Edges are specified between a source node (`src`) and a destination node (`dst`), both referenced by their id. An additional attribute stores the `weight` (or strength) of the connection between the two nodes.

Listing 1.1   A graph with nodes, edges, and attributes

```
 1 {
 2   "nodes": [
 3     { "id": 0,
 4       "label": "Myriel"
 5     },
 6     { "id": 1,
 7       "label": "Napoleon"
 8     },
 9     { "id": 2,
10       "label": "Mlle Baptistine"
11     },
12     // More nodes here ...
13   ],
14   "edges": [
15     { "src": 1,
16       "dst": 0,
17       "weight": 1
18     },
19     { "src": 2,
20       "dst": 0,
21       "weight": 8
22     },
23     // More edges here ...
24   ]
25 }
```

Our listing with three nodes and two edges is only an abbreviated view as indicated by the comments in lines 12 and 23. The graph that we are about to visualize actually contains 77 nodes and 254 edges. It captures the co-occurrence of characters in the chapters of Victor Hugo's *Les Misérables*.

With only the listing of the graph, it is extremely difficult to make sense of the information hidden in the data. Therefore, we (the Who) will now perform a visual analysis to gain insight into the graph. In the first place, we are interested in the structure of the graph (the Why).

A basic method for visualizing graphs is a node-link diagram. Figure 1.1a shows such a simple visual representation of the graph. Nodes are visualized as dots, and edges are represented as links between the dots. There are many different options for laying out the dots on the display. In our case, a force-directed layout algorithm has been applied. As we can see, the structure of the graph, that is, who is connected to whom, becomes quite clear, merely by drawing dots and links.

Let us refine our interest in the data. We are now interested in who are the major characters with the most connections to other characters. We can already extract this information by looking at the number of edges that are connected to a node. Yet, it is a bit cumbersome to count edges, and the clutter of edges makes counting difficult anyway. So how can we make the desired information more readily visible?

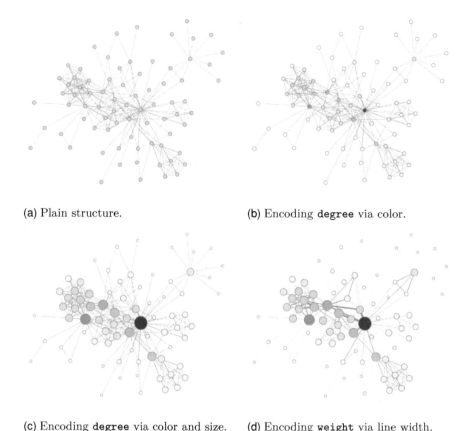

(a) Plain structure.

(b) Encoding **degree** via color.

(c) Encoding **degree** via color and size.

(d) Encoding **weight** via line width.

Figure 1.1 Node-link diagram visualizing a graph's structure and attributes.

The number of edges per node is also called the *node degree*. This derived numeric attribute can be visualized alongside the graph structure. To this end, we assign to each dot a color that represents the node degree. Dark green nodes exhibit a large node degree, while light green nodes have a low degree. Figure 1.1b illustrates the color coding. With this enhanced node-link diagram, we can immediately identify the dark green dot in the center of the figure as the major character in the network.

But we are still not fully satisfied. Low-degree nodes are of the same size as the important high-degree nodes. We want to further accentuate the important characters in the network and attenuate the less relevant supporting characters. Complementing the color coding, we vary the size of the dots depending on the node degree. As can be seen in Figure 1.1c, it is now much easier to assess the importance of characters.

Our visual representation is now quite expressive in terms of information about the nodes of the graph. However, the edges have received only little attention. To render a more complete picture of the data, it would be nice to visualize the edge `weight` as well. This can be achieved by varying the width of the lines connecting the colored dots. In Figure 1.1d, bold lines indicate strong edges, whereas thin lines stand for weak edges. With this additional visual encoding, we can easily see to whom a character is most prominently connected.

In summary, we have now visualized the graph structure and two associated graph attributes. The structure is nicely visible as dots and links. The two attributes node degree and edge weight are encoded visually via color plus size and line width, respectively. The resulting visual representation enables us to *see* the key characteristics of the data. By reading the data's textual representation we could have obtained the same characteristics, but it would have cost us much more time and painstaking brainwork.

1.2.2 Enhancing the Data Analysis

The previous simple examples illustrated the potential of visualization. Yet, simple visual representations alone are often not enough to solve more complex problems. The graph that we visualized consisted of 77 nodes and 254 edges only. However, it is not uncommon to work with graphs with thousands of nodes and edges, and dozens of attributes. Climate networks are an example of such large and complex graphs. They are generated from large-scale simulations of meteorological phenomena with the goal to better understand and predict the development of climatic conditions on earth.

With increasing size and complexity of the data, we have to go beyond the simple visual representations introduced earlier. When visualizing large data, we risk ending up with visual representations that are cluttered. Adequate countermeasures have to be taken. Moreover, it is hardly possible to encode all aspects of the data into a single image. It is rather necessary to provide multiple views on the data, where each view emphasizes a particular data facet. The next two examples illustrate these lines of thought.

Consider the climate network visualized in Figure 1.2a. It contains 6,816 nodes and 116,470 edges. Its visual representation is actually a mess; there are simply too many dots and links. What can we do about this? A standard approach in such situations is to focus on relevant subsets of the data. Subsets can be created dynamically using interactive filtering mechanisms that enable users to specify the parts of the data they are interested in.

For the climate network we may be interested in those nodes that are crucial for the transfer or flow in the network. Such nodes are characterized by a high *centrality*, a graph-theoretic measure. An automatic algorithm can be used to calculate the centrality for each node of the network. Then it is up to the user to determine interactively a suitable threshold for filtering out low-centrality nodes and their incident edges.

Figure 1.2b shows the climate network where nodes with a centrality below 65,000 have been filtered out. As a result the visual representation contains

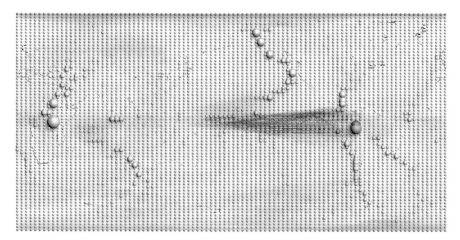

(a) Full graph with 6,816 nodes and 116,470 edges.

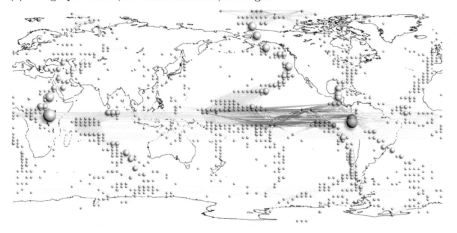

(b) Filtered graph with 938 nodes and 5,324 edges.

Figure 1.2   Dynamic filtering to focus on relevant parts of a climate network.

only those graph elements that are relevant with respect to the interest of the user. Now, there are no more than 938 nodes and 5,324 edges. This obviously reduces clutter and generates a better view on the structures hidden in the data.

With dynamic filtering as described so far, it is possible to deal with problems caused by data size. Another challenge is data complexity. It relates to the many semantic aspects that may be linked to the data. We already mentioned the graph structure and the graph attributes as important aspects. Additionally, there can be spatial and temporal aspects. A climate network is usually given in a spatial frame of reference and it may also be subject to

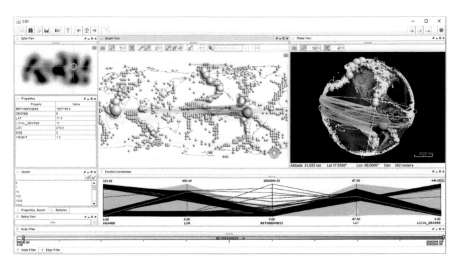

Figure 1.3 Multiple-views visualization of a climate network.

change over time. In order to understand data comprehensively it is necessary to understand the individual aspects and their interplay. This requires multiple dedicated visual representations, each addressing the particularities of a specific aspect. Figure 1.3 depicts a system where multiple views work in concert to visualize our climate network. Without going into too much detail, there are a density plot (top left), a node-link view combined with a map (center), a globe view (right), a multivariate attribute plot (below node-link and globe), a filter slider (bottom), and a few auxiliary controls. All these views are linked. That is, picking a data item in one view will highlight that item in all other views. This linking among views is essential for integrating the different data facets and enabling the user to form a comprehensive understanding of the data.

1.2.3 Considering Advanced Techniques

In the previous paragraphs, we computed graph-theoretic measures, added interactive filtering, and combined multiple linked views to create a comprehensive overview of the data. But how far can we get with interactive visual data analysis? Certainly, there are limits. The visualization has to fit into the available display space. Interaction should not overwhelm users with too many things to carry out manually. Analytic computations have to generate results in a timely fashion.

As we try to push these limits, we have to consider advanced techniques. For the purpose of illustration, we briefly look at two examples. One aims to guide users during the data analysis and the other to expand the screen space for visualization.

It has already been mentioned that interaction is crucial to creative sensemaking. However, interaction can also be demanding. The user has to deal

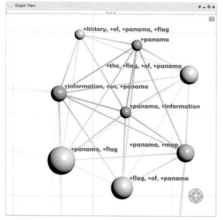 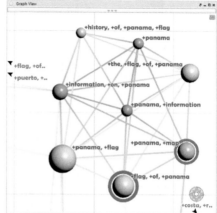

(a) Where should I go next?   (b) Visual cues hint at candidates!

Figure 1.4   Guidance provides assistance during data navigation.

with several questions: What can I do to get closer to my goal, which action sequence do I have to take, how are the individual interactions carried out? An advanced visual analysis system is capable of providing guidance to assist the user in answering such questions.

We illustrate such guidance by the example of another common question: Which part of the data should be visited? A typical approach to study data more closely is to zoom in as shown in Figure 1.4a. Note that we visualize a different graph now, a graph about co-occurrence of search engine query keywords with 2,619 nodes and 29,517 edges. While zoomed in, it is possible to see details, but only for a fraction of the graph. So, the data analysis is an iterative process during which one part of the graph is visited after the other. This iterative process requires users to answer the question where to go next. Should the user be hesitant to continue the navigation of the data, this may indicate that guidance should be provided (the When).

If this is the case, a computational method scans the immediate neighborhood of the currently visible part of the graph in search for nodes or edges that are potentially interesting according to a user-specified degree-of-interest (DoI) function. The visual representation is then enhanced with visual cues that point at the most promising candidates. Figure 1.4b shows a few nodes emphasized with red circles. These nodes are worth investigating in detail. Moreover, arrows at the view border suggest directions in which further interesting nodes can be found. The user is free to follow the given recommendations or to continue the exploration otherwise. Of course, guidance is a delicate means of user support. If guidance is obtrusive, users may not accept it. If it is well-balanced, however, guidance can be a valuable tool.

For our second example of advanced visual data analysis, we are addressing the screen space limit. When pushing this limit, a natural step forward is to

Figure 1.5 Visual data analysis in a multi-display environment. Reprinted from [RLS11].

go for bigger displays. Instead of working in the classic desktop environment, several displays are combined to form so-called multi-display environments (the Where). As shown in Figure 1.5, there are stationary public displays and dynamic private displays, which can enter and leave the environment as needed. Each display may contain several views and the views can be focused and re-arranged to suit the analytic situation at hand.

On the one hand, multi-display environments provide more space for displaying visual representations at high resolution. There can be even multiple users working collaboratively to analyze the data. On the other hand, new challenges need to be tackled. How to distribute views on the displays, how to deal with users occluding the displays, and how to interact with views at the greater scale? Finding answers to these questions and supporting the user in such potent workspaces is part of ongoing visualization research.

In this section, we presented a series of visualization examples. We started with basic visual encodings, incrementally enhanced the visual representations, and finally looked at some advanced techniques. These examples are kind of a teaser of what to expect from this book. The next section will provide some more detailed information on the book structure.

1.3   BOOK OUTLINE

This book is structured into six chapters: the first introductory chapter you are currently reading plus five chapters on interactive visual data analysis to come. Figure 1.6 provides an overview of the chapter structure.

Chapter 2 is concerned with fundamental aspects pertaining to interactive visual data analysis. We will look into design criteria, factors influencing the design, and models describing the involved processes.

Chapter 3 is about visualization. You will learn about the basic methods of visual encoding and presentation, and about various visualization techniques for different types of data.

Chapter 4 is dedicated to interaction. The chapter discusses general interaction concepts and illustrates how interaction techniques can facilitate the visual data analysis in many ways.

Chapter 5 deals with automatic computations to support the visual data analysis. The primary goal will be to reduce the complexity of the data and their visual representations.

Chapter 6 sheds some light on advanced concepts for interactive visual data analysis, including multi-display visualization environments, user guidance, and progressive visual data analysis.

Chapter 7 will briefly summarize the book and outline ideas for readers to continue with the topic of interactive visual data analysis.

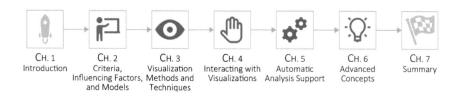

CH. 1	CH. 2	CH. 3	CH. 4	CH. 5	CH. 6	CH. 7
Introduction	Criteria, Influencing Factors, and Models	Visualization Methods and Techniques	Interacting with Visualizations	Automatic Analysis Support	Advanced Concepts	Summary

Figure 1.6   Chapter structure of this book. Icons by `icons8.com`.

FURTHER READING

General Literature: [SM00] ● [Spe07] ● [WGK15]

Criteria, Factors, and Models

INTERACTIVE VISUAL data analysis is highly context-dependent. We will need different techniques for analyzing time-series data than for graph data. We will want to use completely different visual representations for getting an overview of the overall data distribution than for inspecting individual patterns and trends. And we will most likely interact differently when working with data on an interactive surface as compared to a standard desktop.

There is no silver bullet solution that simply scales to all possible analysis scenarios. If there was such a universal approach, there would be no necessity for this book.

The first step to designing context-dependent solutions is to know the fundamental requirements posed by interactive visual analysis scenarios. This aspect will be addressed by introducing corresponding *criteria* in Section 2.1. Second, we need to describe the *influencing factors* that characterize analysis scenarios. This primarily concerns the data to be analyzed and the tasks to be accomplished, but also the people who carry out the analysis and the environment in which it takes place. These influencing factors will be dealt with in detail in Section 2.2. Finally, we need to understand the fundamental processes behind interactive visual data analysis. These will be discussed in Section 2.3, where we cover the design process, the data-transformation process, and the knowledge-generation process. Throughout this chapter, we will use illustrating examples to convey the major points.

2.1 CRITERIA

If you want to be successful in analyzing data with interactive visual tools, you cannot just create an arbitrary visual representation, add some interactivity to it, and spice it with a little computational support. Much of the potential of interactive visual data analysis would be wasted. Even worse, you could end up with findings that are simply not true. Any follow-up decisions based on your analysis would be tainted.

Consider for example the visual representation from Figure 2.1a. The data captures the Germans' life satisfaction in 2017. Satisfaction is measured on a scale from 0 (very dissatisfied) to 10 (very satisfied). The map represents the average satisfaction per region with a red-yellow-green color scale. To the casual observer, Figure 2.1a suggests that people in the eastern parts of Germany are mostly dissatisfied. This is where the visual representation fails, because the visual impression is not backed by the data. A closer look at the legend tells us that the visualized satisfaction is between 6.83 and 7.43, which means people are closer to being satisfied than to being unhappy.

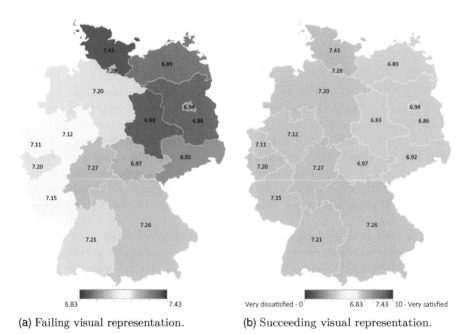

(a) Failing visual representation.   (b) Succeeding visual representation.

Figure 2.1   Visualization of life satisfaction in Germany.

Figure 2.1b shows a successful visualization. The color scale is mapped appropriately to the possible range of values between 0 and 10. The visual representation now suggests that German people are mostly satisfied with their lives. There are only slight differences between the different parts of Germany.

The take-home message of this example is that we always have to ask what makes a good visual representation that actually helps us analyze our data? In fact, we would like to be able to assess the overall *quality* of interactive visual analysis solutions. However, as indicated before, there are many influencing factors, including the properties of the data, the nature of the tasks, the characteristics of the human, the modalities of the output and input system, and the resources of the environment. These factors bear complex questions in themselves and are not easy to define formally. This makes it hard to come up with a coherent definition of quality. Instead, we establish three quality criteria that interactive visual approaches to data analysis should obey.

Expressiveness In the first place, an interactive visual representation must be *expressive*. Expressiveness is a mandatory condition. The assessment of whether an interactive visual representation is expressive or not can only be made in relation to the data to be visualized and the task to be supported. Apparently, we have to consider visual expressiveness and interactive expressiveness separately.

A visual representation is expressive if it communicates the desired information contained in the data, and only this information. In other words, the visual representation neither fabricates nor withholds information, but objectively reflects the information that we need to accomplish our task.

Accordingly, an interactive representation is expressive if it allows us to carry out the actions needed to acquire the desired information, and only these actions. Put differently, the user is enabled to do exactly what is necessary for the task at hand.

Effectiveness Secondly, an interactive visual representation should be *effective*. Effectiveness is a goal-oriented condition. It relates to the degree to which we as humans are able to achieve our analysis task.

An interactive visual representation is effective if it is geared to the human sensory and motor systems, that is, our abilities to observe and interact with our environment. As we are dealing with visual representations, the properties of the human visual system are of primary relevance. In this sense, effectiveness captures how well we can extract the information needed for our task from a visual representation.

In a similar vein, we can characterize effectiveness as a measure of how well we can convey an interaction intent to the computer. Most of the time, this concerns our physical ability to move our fingers, hands, or the whole body with different speed and accuracy, but also commands issued with our vocal tract.

TABLE 2.1   Quality criteria of interactive visual data analysis.

Criterion	Concern
Expressiveness	Faithfully map data and tasks
Effectiveness	Enable users to accomplish task
Efficiency	Balance benefits and costs

Efficiency   Finally, interactive visual analysis should be *efficient*. Efficiency is a desired property. It adds the economic aspect: The gains from using an interactive visual approach should outweigh the computational resources and human effort needed to carry out the analysis.

Resource-wise, we are interested in how much time and how much memory it takes to perform calculations on the data and to transform them into expressive and effective images. The display space needed is an efficiency-relevant resource as well.

On the users' side, effort goes into interpreting visual representations and carrying out interactions. Interpreting visual representations is primarily a mental effort, but also related to physical eye movements. On the other hand, carrying out interactions is mostly a physical effort, but requires mental activities as well for planning and coordinating the interaction.

With expressiveness, effectiveness, and efficiency we have described three criteria that contribute to a notion of quality of interactive visual data analysis. Table 2.1 summarizes the main points of our discussion.

Note that the order in which we introduced the criteria bears an important message: Expressiveness first! No matter how hard we try to be effective and efficient, an interactive visualization is not useful at all if it misrepresents the data. One can also speak of *representational primacy* [AS05]. Already in 1983, Tufte defines this as a fundamental principle of good graphical representations [Tuf83]:

"Above all else show the data."

Tufte, 1983

It is important to realize that the introduced quality criteria are difficult to evaluate formally. While some aspects would be easy to quantify, such as, computation time or display space, others are not. For example, the benefit of a visual data analysis is hard to capture, but we would need it to determine efficiency. Moreover, we deal with aspects of the human user, whose cognitive processes are not yet fully understood and whose goals and tasks are hard to define on a formal level.

Nonetheless, the described quality criteria provide us with a basis for critically questioning the visual representations, the provided interactions, and the involved calculations and transformations.

2.2 INFLUENCING FACTORS

In order to build interactive visual analysis solutions that are expressive, effective, and efficient, we must take a closer look at the factors that influence the analysis. These factors pertain to the *subject* of the analysis, that is, the data, to the *objective* of the analysis, that is, the goals and tasks, and to the *context* of the analysis, that is, the human and technical resources involved.

2.2.1 The Subject: Data

Data on a computer are but sequences of zeros and ones. They are worth nothing without knowledge about how to decipher them. Consequently, if we want to make sense of data, we need both the data themselves plus a description of how to interpret them. This description of data properties is the first factor to be taken into account for choosing or building appropriate analysis solutions.

Data Domain

So, what characterizes an individual datum (or data value)? Any data value comes from a *data domain*. The data domain is the set of values that can potentially appear in the data.

An important property of a data domain is its *scale*. The scale determines what relations and operations are possible for the data values in the domain. At the top level, we can differentiate *qualitative* (or categorical) and *quantitative* (or numerical) data. At a second level, we can further categorize qualitative data into *nominal* and *ordinal* data, and quantitative data into *discrete* and *continuous* data. The different data scales and the relations and operations they permit are summarized in Table 2.2. Next, we look a bit closer at these different data.

TABLE 2.2 Operations possible in different data domains.

	Qualitative		Quantitative	
	Nominal	Ordinal	Discrete	Continuous
Equality	•			
Order	•	•		
Distance	•	•	•	
Interpolation	•	•	•	•

Nominal Data For nominal data, we can assume the existence of an equality relation =, which allows us to determine whether two values are equal. This is the most primitive insight that can be gained about data values. Additionally, we can count frequencies of nominal values.

An example of nominal data would be identifiers such as names {Anika, Ilka, Maika, Tilo, ... }.

Ordinal Data For ordinal data, an order relation < exists in addition to the equality relation. This allows us to determine whether a data value is smaller (or less, before, weaker, of lower rank) than another data value. With the help of the order relation, it is possible to sort or rank data values.

An example of ordinal data would be age groups such as {children, youths, adults, elders}. With the order relation, we can say that children < adults.

Nominal and ordinal data do not have an inherent notion of distances between data values. This changes with quantitative data, which are based on metric scales as we will see next.

Discrete Data Discrete data are numeric data whose domain can be equated to the set of whole numbers $\mathbb{Z}$. This implies that we can count discrete values and determine the difference between any two data values by means of a distance function.

An example of discrete data would be the number of people visiting a doctor. If one day 34 people seek treatment and another day 23, we can naturally derive a difference of 11 persons.

Continuous Data Continuous data are numeric data whose domain can be equated to the set of real numbers $\mathbb{R}$. As such, continuous data are uncountable. We can also say they are dense, that is, between any two data values, a third data value exists. This property is a necessary condition for being able to carry out interpolations on the data.

An example of continuous data would be temperature values as measured hourly by a weather station. From two data values 10.6°C and 13.2°C measured at 8:00 and 9:00 we may interpolate 11.9°C for 8:30 (in the case of stable weather conditions).

Note that it is not always obvious to which category a data value belongs. For example, when dealing with customer numbers or zip codes, we may think they are discrete data. However, they are actually categorical identifiers, for which neither order nor distance have a meaningful interpretation.

Later in this book, we will see that the different data scales require different visual encodings. For example, for ordinal data, the ordering of data values must be clearly communicated, yet without suggesting any notion of distances between data values. More details on how to appropriately visualize data depending on their scale will be discussed in Chapter 3.

With the scale of data domains, we can characterize individual data values. If multiple numerical data values are combined, we can further distinguish scalar data, vector data, and tensor data.

If we are dealing with a single numeric value, we speak of a *scalar*. A scalar datum contains but the value itself, such as the temperature values mentioned before. Multiple numeric values can be combined to form a *vector*. A vector datum defines a direction plus a magnitude based on the vector components. An example is velocity vectors to describe motion in physics.

Still more information is captured by a *tensor*. A tensor datum consists of multiple directions and magnitudes. The order of a tensor defines how much information it bears. Scalars and vectors are special cases of tensors with order 0 and 1, respectively. A 2nd-order tensor can be represented by a matrix, tensors of higher order by multi-dimensional arrays. An example of tensor data are 2nd-order stress tensors whose 3×3 components describe the stress at a point inside deformed material.

In this book, we focus on scalar data. The interactive visual analysis of vector data and tensor data is a challenging problem on its own. The interested reader is referred to books specialized on these topics [HJ05; Tel14].

Data Structure

In the previous paragraphs, we introduced data domains to characterize an individual piece of data. However, an interactive visual analysis of a single piece of data does not make much sense. We are interested in analyzing entire sets of data. In reality, data are often unstructured and contain various chunks of heterogeneous information. The analysis of such messy data is extremely difficult. Therefore, data should ideally be stored in or be transferred to structured formats.

One such format that is universally applicable is the *data table*. A data table consists of rows and columns. The columns represent data *variables*. Each variable is associated with a *data domain* that specifies the values that can *possibly* appear in a column. The values that *actually* do appear in a column define the *value range*.

The rows of a data table represent data *tuples*. A tuple consists of a set of data values and can be understood as a unit of data that describes the properties of an observed entity. There is one value for each variable, and the value is from the variable's domain. Depending on the context of use, tuples are also called observations, records, items, or objects. We will often use the more general term *data element*.

Particularly in scenarios where relations exist between data entities, these relations may define another layer of structural organization. *Hierarchies* are common to structure data in a top-down fashion according to different levels of detail (or abstraction). More generally, data can be modeled as *graphs* whose nodes and edges represent data entities and the relations among them, respectively. Examples of graph-structured data are social networks or biochemical reaction networks. Both hierarchies and graphs can be stored by using two data tables, one for the entities and one for the relations.

Another option to store data in a structured manner is to use data *grids*. Data grids are particularly important in the realm of flow visualization and volume visualization. As these topics are not in the scope of this book, we again refer the reader to relevant specialized literature [HJ05; PB07; Tel14].

Data Space

A data table as introduced before serves to structure sets of data values. However, structure alone is not sufficient. We additionally need to consider the characteristics of the *data space* that is spanned by the variables.

An important point is the distinction between *independent* and *dependent* variables. Independent variables correspond to the *dimensions* of the space where the data have been collected, observed, or simulated. In turn, the dependent variables describe the *attributes* of what has been collected, observed, or simulated. More formally, we may model this aspect as a functional dependency as follows:

$$f : (D_1 \times D_2 \times \cdots \times D_n) \to (A_1 \times A_2 \times \cdots \times A_m)$$

where D_i denote the dimensions (independent variables) and A_i the attributes (dependent variables). The definition of f implies that a point in the *reference space* is associated with exactly one data point in the *attribute space*. A schematic depiction is given in Figure 2.2.

Figure 2.2 Functional dependency between the reference space and the attribute space. For a point in the reference space, there is exactly one point in the attribute space.

It is typically this functional dependency that has to be made visible and understood during interactive visual data analysis. Assume, for example, your data consist of temperature and air pressure observed at different locations and times: $(latitude \times longitude \times time) \to (temperature \times pressure)$. You might be interested in studying how the measured attributes vary over time or where extrema are located. In order to support these and similar analysis questions, it is necessary to visualize the functional dependency between dimensions and attributes faithfully. Note that there are also analysis objectives for which the functional dependency plays only a minor role, for example, when acquiring a general overview of the data's value ranges.

As a graphical summary, Figure 2.3 collects the key terms that we have discussed so far.

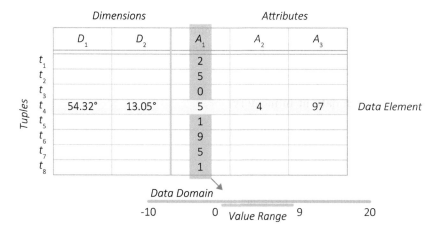

Figure 2.3 Key terms for characterizing data.

Data Size

With the help of a tabular model of data and the distinction of dimensions and attributes, we can now study another important characteristic of data: their size. To this end, let's take a look at a data table with dimensions, attributes, and tuples as illustrated in Table 2.3. We can see that the size of a dataset is determined by n, the number of dimensions, by m, the number of attributes, and by k, the number of tuples.

TABLE 2.3 A data table.

D_1	D_2	$\cdots$	D_n	A_1	A_2	$\cdots$	A_m
$d_{1,1}$	$d_{1,2}$	$\cdots$	$d_{1,n}$	$a_{1,1}$	$a_{1,2}$	$\cdots$	$a_{1,m}$
$\vdots$							$\vdots$
$d_{k,1}$	$d_{k,2}$	$\cdots$	$d_{k,n}$	$a_{k,1}$	$a_{k,2}$	$\cdots$	$a_{k,m}$

In general, we can say we are dealing with k tuples of n-dimensional m-variate data. Data that consist of more than one dimension are denoted as *multi-dimensional* data. If there are many dimensions, the term *high-dimensional* data is common. A similar notation is used for the attributes. If there is only one or two attributes, we use the terms *univariate* and *bivariate* data, respectively. The presence of several attributes is indicated by the term *multivariate* data.

Obviously, the larger m, n, and k are, the more challenging and complex the interactive visual analysis will be and the more sophisticated tools will be needed to support the analysis.

The question of whether a dataset can be regarded as *large* or not is often discussed controversially and typically cannot be answered unambiguously.

One way to resolve this issue is to think of the different bottlenecks that data have to pass on their way from the computer to the human mind. This can be made clear by asking the following questions:

- How much data fit in storage?

- How much data fit in memory?

- How much data fit on the display?

- How much data can we digest?

If for any of these questions our data do exceed the corresponding limit, we may consider them *large*. The earlier a limit is exceeded in this list of bottlenecks, the *larger* the data typically are. In any case, we will need additional methods, such as interaction or computational support, to overcome the bottlenecks. These topics will be studied in Chapters 4 and 5.

So far, we have discussed data properties that can, to a certain degree, be extracted from the data. We can look at the values and make assumptions about the associated data domains. We can inspect the variables and determine which are independent and dependent. And of course, we are able to count the variables and tuples of a dataset to get its size. Next, we will deal with a property that cannot be derived, but must be provided to us: the data scope.

Data Scope

In order to create expressive visual representations of data, we need to know the data's *scope*. The data scope characterizes the validity of data at or around the point of observation. We can distinguish between three types of scopes as illustrated in Figure 2.4:

- **Global Scope:** Data are valid across the entire reference space.

- **Local Scope:** Data are valid at a point of reference and its vicinity.

- **Point Scope:** Data are valid only at a point of reference.

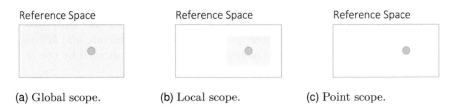

(a) Global scope.   (b) Local scope.   (c) Point scope.

Figure 2.4 The scope defines to which extent an observation is valid.

Whether individual values are valid globally, locally, or point-wise cannot be determined from a dataset per se. Instead a data description is needed, which should include hints as to which scope is applicable.

For specific applications, we may assume a certain data scope. One such example is geo-spatial data, which, according to Tobler's first law, have a local scope, because attributes measured at proximal locations tend to be correlated [Tob70]. Unfortunately, the local scope is not always defined precisely. From a visualization perspective, the question is how the local validity of data can be represented?

Let us illustrate this with the three different representations of measurements of water quality as shown in Figure 2.5. The data are color-coded with a green-to-red color scale, where green represents high quality and red signals low quality. Figure 2.5a neglects the local scope and shows us just the data points. This kind of representation is rather sparse and does not really support building a good understanding of the data.

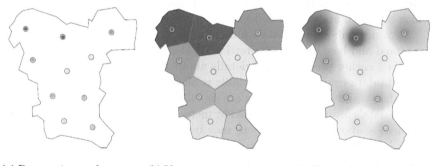

(a) Data points only.   (b) Voronoi partitioning.   (c) Shepard interpolation.

Figure 2.5   Visualizing the local scope of measurements of water quality.

In Figures 2.5b and 2.5c, interpolation comes to our help. By interpolation, we can assign a color to all pixels in order to create a dense visual representation that communicates the local scope of the data. Figure 2.5b uses nearest-neighbor interpolation, which corresponds to a partitioning of the space into discrete Voronoi regions. Now we get a much better impression of the water quality. However, the discrete regions suggest discontinuities at their borders. Certainly, no such discontinuities exist in the water. In Figure 2.5c, we used Shepard's interpolation method [She68]. The visual representation is smooth, which more closely corresponds to what we expect from the data.

Note, however, that interpolating scattered data is a non-trivial problem. There are various methods and each comes with its own set of parameters and corresponding results. Which methods and parameterizations are appropriate must be decided on a case-by-case basis.

Meta-data

All of the previously discussed properties of data (domain, structure, space, size, and scope) should be described and provided along with the actual data. This data description is also called *meta-data*, or data about data. Ideally, the meta-data are curated with the same scrutiny as the data themselves, because the meta-data are the first to be consulted for an informed decision on how to visually analyze the actual data.

Moreover, meta-data can (and should) contain additional information about the evolution of the data. This includes the data's past, the data's present, and the data's future [Sch+17].

In terms of the data's past, we may think of information about how the data have been collected, observed, or simulated. This so-called *data provenance* information can be quite important when it comes to understanding and replicating the results from analytic activities. An example of provenance meta-data is indicators of *data quality* to specify how much confidence we may have in the recorded data. A closely related notion is that of *data uncertainty*, which tells us if and to which degree the data may be uncertain.

Meta-data about the data's present state basically specify how the data are stored and how they can be retrieved. This covers all the data descriptors described earlier plus technical details about the *data format*. For example, the meta-data can tell us if a variable is allowed to include *missing data* (or null values), and if yes, how they are signaled.

The data's future may be covered by meta-data as well. This mainly includes information about what operations may be performed on the data. Such *data utility* information is somewhat rarer and often not given explicitly. But it can be quite valuable. For example, we already mentioned data interpolation as difficult to deal with. Meta-data can give us information about whether interpolation is possible and under which assumptions.

In summary, meta-data are as important as the data to be analyzed. Figure 2.6 provides an overview of the various aspects discussed in the previous paragraphs. While our considerations remained rather theoretical so far, the next section will be more practical by describing common classes of data.

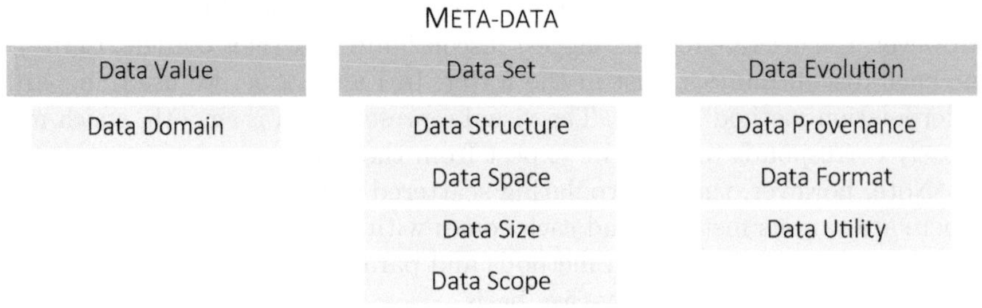

META-DATA		
Data Value	Data Set	Data Evolution
Data Domain	Data Structure	Data Provenance
	Data Space	Data Format
	Data Size	Data Utility
	Data Scope	

Figure 2.6 Meta-data to characterize the data to be analyzed.

Classes of Data

Let us next take a look at classes of data that are common in practice. We introduce the following abstract notation: A stands for (one or more) data *attributes*, T for *time*, and S for *space*, and R for structural *relationships* among the data. Depending on which of these aspects are present, different classes of data can be distinguished. Figure 2.7 provides an overview. More details and example data are described below, where the arrow symbol → is used to indicate a functional dependency between reference space and attribute space.

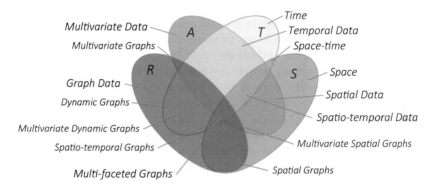

Figure 2.7   Four-set Venn diagram illustrating different classes of data.

Multivariate Data Multivariate data consist of several data attributes A. Typically, the data analysis will concern the distribution of data values and correlations among the attributes. An example is player statistics as maintained in many leagues of various sports.

Temporal Data For temporal data, the reference space is defined by the dimension of time, and one or more time-dependent attributes are observed over time $T \to A$. The data analysis typically seeks to understand how the attributes evolve over time. This includes the detection of trends or cyclic behavior. Stock prices are a common example of temporal data.

Spatial Data If data are given in a spatial frame of reference, we denote them as spatial data $S \to A$. Two-dimensional and three-dimensional spatial data are common. The analysis of spatial data is centered around finding spatial patterns, clusters, or hot-spots. An example of spatial data could be the distribution and richness of mineral deposits.

Spatio-temporal Data Quite often, spatial data do also include a temporal dependency, in which case, we speak of spatio-temporal data $S \times T \to A$. An analysis of such data combines spatial and temporal issues, such as the development of spatial clusters over time or cyclic re-occurrence of hot-spots. Spatio-temporal data are ubiquitous in the form of weather data.

Graph Data Graphs consist of a set of data entities, the nodes, and a set of relations between the entities, the edges. In the first place, we are interested in understanding the structural information R inherent in graphs. Additionally, nodes and edges of a graph can also be associated with data attributes to be analyzed $R \to A$. Graphs can even be linked to a spatial context or vary over time, in which case different dependencies could be defined, such as $R \to S \times T \times A$ or $S \times T \to R \times A$. An example of so-called *multi-faceted graphs* is climate networks, which are employed to simulate and predict climatic phenomena on earth.

The introduced classes of data are deliberately kept general. This allows us to map practical analysis problems to these classes. For example, the analysis of documents or document collections could be done by modeling documents as attribute vectors among which relations exist $R \to A$. The analysis of medical images naturally maps to $S \to A$, where S corresponds to the pixel grid and A is the pixel color. In a similar way, the analysis of flow data maps to $S \to A$ (or $S \times T \to A$, if the data are time-varying), where S is a 2D or 3D grid-structured space and A defines the components of vectors.

If the visual analysis is about more complex artifacts such as software systems or distributed processes, it is typically necessary to break them down into smaller conceptual pieces before a meaningful mapping to A, T, S, and R is possible.

With the introduction of general data classes, we end our discussion of data characteristics as the first influencing factor to be considered. Next, we continue with analysis tasks as the second influencing factor.

2.2.2 The Objective: Analysis Tasks

In the previous section, we investigated the subject of interactive visual analysis, the data. In this section, we study the objective, that is, the analysis tasks. Tasks are important particularly for two reasons. First, they determine which portions of the data are relevant for the analysis, and second, they are crucial for meeting the effectiveness criterion as introduced in Section 2.2.

Let us illustrate this with an example. Figure 2.8 shows two color-coded maps with identical geo-spatial data. Figure 2.8a uses a yellowish-to-green color scale where color varies in perceptually uniform steps to support the task of *identifying* the data values. The map in Figure 2.8b uses a color scale that emphasizes extrema, whereas intermediate values are attenuated by the use of gray tones. This encoding facilitates the task of *locating* minima (in blue) and maxima (in red) on the map. Our example clearly demonstrates that the generation of effective visual representations depends on the task at hand.

In fact, we can state that a particular visual representation supports particular tasks: Data + visual representation $\mapsto$ tasks. In turn, this implies that specific tasks require dedicated visual representations: Data + task $\mapsto$ visual representation. The very same dependency on the task exists for the

(a) Coloring suited to identifying values.   (b) Coloring suited to locating extrema.

Figure 2.8   Different visual encodings to support different tasks.

interaction design and the involved analytical computations. As we see, it is the task that determines which design, technique, or method would be appropriate for an interactive visual analysis of data.

However, tasks are notoriously difficult to conceptualize due to their multi-faceted character. A task is usually carried out to achieve a larger *goal*. Moreover, a task is typically focused on an *analytic question* that operates on a relevant part of the data, the task's *target*. Finally, a task can be accomplished by employing different *means*. In the following, we will describe in more detail the goals, analytic questions, targets, and means that characterize tasks.

Goals

Goals describe the intent with which analysis tasks are pursued. Goals have an overarching character in that achieving a goal typically involves several steps of analytic activity. General goals are to *explore, describe, explain, confirm,* or *present* the data [Sch+13a; LTM18].

Exploration is geared towards making first observations, such as discovering trends in heterogeneous data or detecting outliers in homogeneous data. This also includes noticing the absence of something that was actually expected in the data such as a commonly known pattern or trend that is not there. In a sense, exploration follows an *I-know-it-when-I-see-it* approach to *undirected* search, or as pioneers of visual analytics put it [TC05]:

"[...] detect the expected and discover the unexpected."

Thomas and Cook, 2005

Description is all about characterizing an observation by the associated data elements, and thereby deriving a specification for an observation. For example, an outlier can be described by its characteristic values and, if available, its spatio-temporal context. A proper description may serve as a basis for configuring further analysis steps. In particular, a description allows for sharing first insights with other people, who can later be involved in verifying the analysis results.

Explanation means identifying all contributing data and finding the main causes behind an observation. This involves investigating several questions. Is the observation by itself significant or did we just interpret too much into the noise among the data? Does the observation re-occur throughout the data or are we looking at a singular outlier produced by unlikely circumstances? If the observation does re-occur, does it show up reliably under the same conditions, thus forming a pattern, or are its appearances seemingly random?

Such investigations are necessary to understand the expected and recognize the reasons behind the unexpected. By trying to explain observations, we will obtain more and deeper insights, which will lead to hypotheses about the data.

Confirmation aims to verify the hypotheses. Unlike exploration, confirmation is a *directed* search. We look for something concrete, some evidence to either back up or refute a hypothesis in terms of validity, generality, or reliability. For this purpose, we may study alternative visual representations of the data or re-parameterize the analytic computations in order to check whether this leads to the same results. If available, related data may be consulted to test if they exhibit similar results.

Presentation is to communicate confirmed analysis results. While explanation and confirmation were about convincing ourselves, presentation is about convincing others of what we have found in the data. This is best done by telling a story about the data, the analysis, and the results. Such a story can act at different levels of emphasis. We may *inform* an audience by letting the results speak for themselves, *explicate* the results to an audience, or even *persuade* an audience into agreement with the results. The audience in this context can be the listeners of a talk, the readers of an article, or colleagues participating in a scientific discussion.

Exploration, description, explanation, confirmation, and presentation can be understood as subsequent phases in the data analysis process. We start exploring the data until we make an observation. Next we describe the finding and try to explain it. Then we confirm that our hypotheses about the data are valid. Finally, we are ready to present the confirmed analysis results. Note how this workflow proceeds with increasing specificity from not knowing that certain observations can be made in the data to knowing them well enough to present them.

The previous paragraphs outlined in broad strokes different goals of analytic activities, yet without detailing what is actually done. This is where concrete analytic questions enter the stage.

Analytic Questions

Analytic questions describe what specifically is sought in a particular analysis step. There are many different questions that are relevant in data analysis scenarios [KK17]. In general, we can distinguish between two fundamental categories: *elementary* and *synoptic* questions [AA06].

Elementary Questions   Elementary questions refer to individual data elements. This includes questions that concern a single data element, but also multiple data elements. What is essential is that each data element is studied individually. Elementary questions may be concerned with:

- Identify: What is the value?

- Locate: Where is the value?

- Compare: Is it less or more?

- Rank: Is there any order?

- Connect: Are they related?

- Distinguish: What makes the difference?

Synoptic Questions   Synoptic questions refer to groups of data. As such, synoptic questions are concerned with the characteristics of sets of data elements, rather than individual data elements. The previously listed questions can operate on groups as well. Additionally, synoptic questions may ask:

- Group: Do they belong together?

- Correlate: Are there any dependencies?

- Trends: Do they develop systematically?

- Cycles: Do they re-occur periodically?

- Outliers: Are they special with respect to the rest?

- Features: What is characteristic for the data?

Note that the two lists of analytic questions are not comprehensive. One can easily imagine many more questions that could be asked about data. The further readings collected at the end of this chapter provide additional information in this regard.

To recap, analytic questions specify what we would like to know about the data. However, as indicated by the unspecific word *they* in the above lists, a question does not tell us where to look. This aspect is covered by the target of tasks as described next.

Targets

Typically, not all data are equally important in the context of a specific analytical question. The target of a task is about *where* in the data a task actually operates. But why is this important?

Specifying a task's target enables us to narrow down what we need to look at to accomplish the task. Are we interested in the data values just by themselves, or how they distribute in a particular spatial or temporal context? Do we want to identify individual data values of the whole data set, or only with respect to a certain subset? Do we search for global trends, or are we interested in more fine-grained variations?

Data of Interest   More specifically, the target defines the *data of interest*, that is, the subset of the data that is indeed relevant to the task. Conceptually, such a subset can be created with respect to the data variables and the data elements. A target can cover all variables of the data space or only a particular subset of variables. Restricting the view on particular variables is called data *projection*.

The data of interest can be narrowed down further by focusing on particular data elements. To this end, we may specify criteria that data elements must exhibit in order to be considered relevant to the analysis. This way it is possible to restrict the data of interest to specific value ranges. Restricting the view on particular data elements is called data *selection*.

Figure 2.9 illustrates how projection and selection constitute the target. Note that the target is not necessarily a closed chunk of data as illustrated in the figure. The general case is that the target is made up of several pieces of data that are distributed across the data space.

Looking at Figure 2.9 we can see that the target is much smaller than the entire data. It can even be as small as a single data element, for example, when working on elementary questions such as identifying a value in the reference space. If the target is a proper subset of the data, elementary or synoptic questions could be addressed depending on whether the data elements are considered individually or as self-contained groups. A target that refers to all data elements typically supports questions of synoptic character.

Data Granularity   For the case that the data are structured hierarchically according to different levels of detail (or abstraction), a target is further characterized by its *granularity*. The granularity defines the level of abstraction needed to accomplish a task.

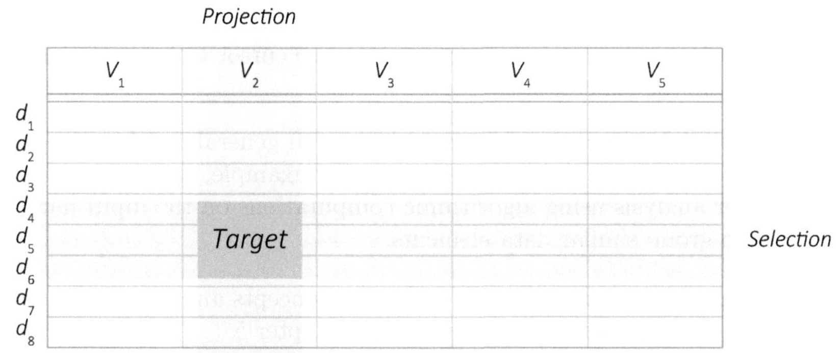

Projection

	V_1	V_2	V_3	V_4	V_5
d_1					
d_2					
d_3					
d_4					
d_5		Target			
d_6					
d_7					
d_8					

Selection

Figure 2.9   A target as defined by projection and selection.

Carrying out tasks at a low level of detail (high level of abstraction) is suited to get a general overview of the data. At this coarse-grained level, tasks are concerned with investigating fundamental data features such as general correlations or overall trends.

On the other hand, tasks can operate on a high level of detail (low level of abstraction). Fine-granular targets typically permit insights into details that cannot be seen in an overview. An example would be small fluctuations in an otherwise monotonically developing trend.

We know now how targets define where in the data a task operates in terms of the data of interest that are studied at a particular data granularity. Next, we briefly describe the means that can be employed to carry out tasks.

Means

There are many options for carrying out the tasks. For example, if we need to locate a particular data element whose specifics are known, we can *query* the data directly. However, if the data element is only vaguely specified, formulating a query is hardly possible. In this case, we have to *scan* the visual representation in order to find it. As we see, querying and scanning are both suitable means to accomplish the same task.

In general, the means describe *how* a task is performed. From a conceptual point of view, one can differentiate between visual, interactive, and computational means.

Visual Means subsume all kinds of visual inspection. The task is carried out with the eyes, in fact, with the entire human visual system. For example, we might visually identify a data value by looking up a color in the visual representation's color legend.

Interactive Means relate to interactive information gathering. In this case, tasks are carried out by the hands or other parts of the human motor system. To continue the previous example, a data value could also be identified by hovering over it with the mouse cursor to bring up a label that shows the exact value.

Computational Means stand for calculations in general. In this case, the computer produces the desired results. For example, we could perform a cluster analysis using algorithmic computations on the input data in order to group similar data elements.

Each category of means is based on its own concepts and methods, which will be detailed in Chapter 3, Chapter 4, and Chapter 5.

Nonetheless, visual, interactive, and computational means are usually applied in concert to fully exploit the cognitive capabilities of the human and the computational power of the machine. Let's illustrate this with the example of finding outliers. We might start with an initial visual inspection of the data followed by an interactive selection of relevant data. Then, a computation of distances could relate the selected data to other data elements in order to detect outliers. Finally, we may again want to do a visual inspection to examine whether outliers have been flagged correctly.

In this section, we have characterized analysis tasks with respect to goals, analytic questions, targets, and means, which indicate *why* a task is pursued, *what* a task seeks, *where* a task operates in the data, and *how* a task is actually carried out. Figure 2.10 collects the key terms that we dealt with. They allow us to formulate tasks like *describe groups of data elements with low values by marking them interactively*. Two more examples are given in Table 2.4. Such task descriptions can then be used to inform the design or support the selection of interactive visual analysis solutions [Sch+13a].

In conclusion, the characteristics of the tasks and the properties of the data are two important influencing factors. Together, they form the basis for some of the most influential early classifications of visual approaches to data analysis [KK93; Shn96]. However, data and task are not the only factors to be observed as we will see in the next section.

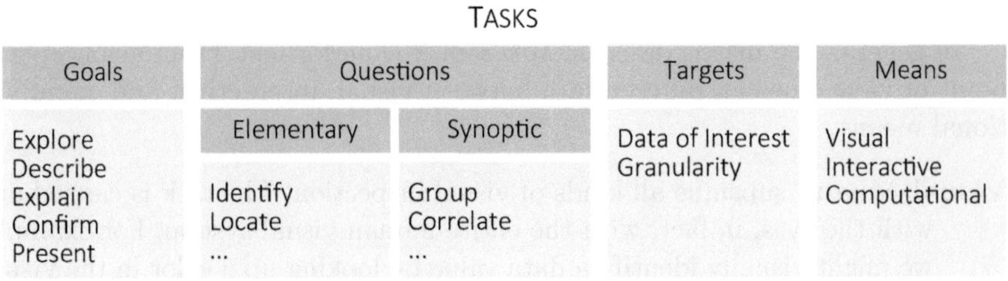

TASKS

Goals	Questions		Targets	Means
Explore Describe Explain Confirm Present	Elementary Identify Locate ...	Synoptic Group Correlate ...	Data of Interest Granularity	Visual Interactive Computational

Figure 2.10 Goals, questions, targets, and means characterize analysis tasks.

TABLE 2.4 Examples of tasks.

Goal	Question	Target	Means
Explore	locations of	maximum values	visually
Describe	groups of	low-value elements	by marking
Confirm	cyclic behavior	of temperature	by statistics

2.2.3 The Context: Users and Technologies

To enable an effective and efficient interactive visual data analysis, the context of the analysis has to be considered as the third important influencing factor. Primarily, we have to ask *who* is analyzing the data and *where* does the analysis take place. In this sense, the context subsumes aspects of the *users* who conduct the analysis and aspects of the *technologies* that constitute the analysis environment.

It is beyond the scope of this section to comprehensively characterize human users and different technologies. Instead, we will briefly look into the key concerns that need to be taken into account.

Human Users

Concerning the question of *who* is carrying out the data analysis, the following aspects are relevant:

Human Factors An important factor is the properties of human individuals. What are we able to see with our visual system and what are we able to do with our motor system? Interactive visual analysis solutions should be centered around the abilities of humans in general. For example, when interacting at larger displays via touch gestures, our precision is usually limited for the majority of users.

Moreover, no user is like the other. Therefore, it is important to consider not only general characteristics of humans, but also the properties of individual users. For example, we need to adapt visual encodings for people with color vision deficiencies so as to allow them to gain the same insight as people with normal vision.

User Background and Expertise People who work with interactive visual analysis solutions come from different backgrounds and have different expertises.

Visual analysis experts excel in creating interactive visual analysis solutions. They apply their tools routinely to examine various kinds of data. However, they are usually not trained in the field to which the analysis is supposed to bring new insights.

On the other hand, *domain experts* are focused on application-specific problems and data, of which they often have a quite good mental model.

They expect analysis tools to adapt to their workflows, and not the other way around. Domain experts could prefer specific representations or ways of doing things, simply because they are widely used in their application background, even if other alternatives would be better suited.

Casual users from the general public will typically favor basic interactive charts and maps over more sophisticated visual analysis systems. Therefore, casual users need to be provided with visual representations that are easy to interpret and interactions that are easy to carry out. It can even be necessary to include incentives and guidance to motivate people to take a closer look and to assist them in doing so.

Application Domain Only if we have a sufficient understanding of the problems that need to be solved in an application domain can we develop suitable analysis solutions. In addition to addressing domain problems, we also need to consider domain conventions. For example, the U.S. Geological Survey suggests using specific colors for geologic maps. As a consequence, these colors cannot be employed for other visualization purposes than geologic characteristics in this particular context.

Application domains can also pose hard constraints on the analysis system. For example, medical diagnosis systems certainly must be super reliable and heavily tested before they can be used in clinical practice.

Single-user and Collaborative Analysis Data analyses can be carried out by a single user or by multiple collaborating users. In the former case, the applied means can be tailored to the needs and preferences of a single individual. In collaborative settings, however, we need to consider a larger pool of methods that additionally need to be flexibly parameterizable in order to be able to attune them to a broader variety of user needs. Moreover, while single-user analysis can focus on the exchange of information between the user and the machine, collaborative analysis further needs to consider the interaction between users, for example, for sharing visual representations and discussing findings.

Technology

To address the question of *where* the data analysis is performed, the following technology-oriented aspects are relevant:

Computational Resources Analyzing data with the help of computers implies the use of computational resources. In the case of interactive visual data analysis, we need resources to create and render visual representations, ideally in high quality and at interactive frame rates. For larger data, we need additional computational resources to carry out analytical pre-processing steps.

Depending on the computational resources of the environment and in line with the actual analysis objective, it must be decided how to balance between the different demands. For example, when users interact, immediate visual feedback is crucial. So, visualization quality could be sacrificed temporarily in favor of quicker response rates during interaction. Similar trade-offs are necessary in many analysis situations.

Display Technologies Nowadays, there exist a variety of technologies for presenting data. It makes a big difference to visualize data on a small portrait-oriented display of a mobile phone, a large landscape-oriented mega-pixel display wall, or a professional designer monitor with high contrast and a wide color gamut.

In general, display technologies are characterized by their physical size, their aspect ratio, their resolution, and their ability to reproduce colors faithfully. The physical size largely determines if a display is applicable in particular environments. The available pixel resolution indicates how much data can be represented. The display's aspect ratio influences the layout of the presented data. The available colors naturally have an impact on the use of colors in a visualization.

Input Modalities An analysis environment is also characterized by its input modalities. They decide about how the interaction is carried out and how precisely an input can be made. Classic mouse and keyboard devices are common in regular desktop scenarios.

Modern displays facilitate interaction via touch technology. This has the advantage that the interaction can take place exactly where the data are shown, that is, on the display. A disadvantage, though, is that smaller graphical objects are more difficult to pick accurately.

In the case of large display walls, mouse and touch interaction alone are impractical due to their limited reach. Alternatives are to track the user's position and gaze and utilize the tracking information to steer the visual data analysis.

As we see, a variety of aspects characterize the context of interactive visual data analysis. Although we have described them one by one, the different aspects exhibit numerous interdependencies. For example, visualizing larger data on a display wall implies that we also have to think of the increased resources that we need to render the visualization and of the interaction modalities that will allow a human user to operate in a large-display setting ergonomically. For another example, if we intend to develop a color-intensive visual representation, we must make sure that it is used on displays with a wide color gamut, and we should also think of ways to circumvent the problem of defective color vision.

In summary, we have now described the data, the tasks, and the context as the major influencing factors. Together, they not only inform the design of interactive visual analysis solutions, but they also determine whether a solution meets the criteria defined in Section 2.1, that is, how well and how fast users can interprete their data, and how balanced the use of resources is.

The condensed take-home message of this section is: If you want an expressive, effective, and efficient analysis solution, you first need to know what data are to be studied, what tasks are to be carried out, and what the context is in which the analysis is conducted. The more comprehensive the answers to these questions are, the easier it will be to come up with an appropriate solution.

2.2.4 Demonstrating Example

In the following, we will demonstrate how the data, the tasks, and the context may manifest in practice. In order to keep our example manageable, we will briefly characterize the data and the context, whereas the tasks will be described in greater detail.

The Data and the Context   For our example, we will use a meteorological time series, which conceptually is an instance of temporal data $T \rightarrow A$. The data are multivariate in that they comprise several attributes, including air temperature, air pressure, wind speed, hours of sunshine, cloud cover, precipitation, and precipitation type. The individual values are scalars with continuous, discrete and nominal scale. For instance, air temperature is continuous, hours of sunshine is discrete, and precipitation type is nominal. We will be investigating more than 23,725 daily measurements, which amounts to about 65 years worth of data.

For the sake of simplicity, we will assume the data analysis is carried out by a single user and takes place on a standard desktop computer with a regular display and the typical mouse and keyboard input. Although our example has been supported by an expert from climate impact research, we do not include any application-specific conventions or constraints. This is again to keep the example simple.

The Task   Now, let's assume our *goal* is to *explore* the data in search of something interesting. More specifically, we are interested if there are any patterns in the data corresponding to seasonal variations, which means, we are dealing with a *synoptic analysis question* for which we need all data elements. Nonetheless, in order to narrow down our analysis problem somewhat, we begin by concentrating on the three attributes hours of sunshine, air temperature, and cloud cover as the *target*.

The charts in Figure 2.11 suggest that we will be accomplishing our task with the help of *visual means*. There is one chart per targeted attribute. Each chart shows time along the horizontal axis and a time-dependent attribute along the vertical axis. For each pair of date and measurement, a dot is placed in the chart.

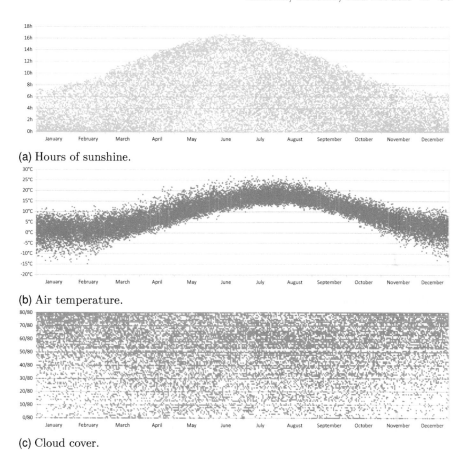

(a) Hours of sunshine.

(b) Air temperature.

(c) Cloud cover.

Figure 2.11   Meteorological measurements over the course of the year.

Note that the time axis spans only the days of the year from January 1st on the left to December 31st on the right. In fact, only day and month of a date determine where to place a dot horizontally, the year is ignored. Conceptually, this corresponds to a *projection* with the effect of mapping the data of several years onto a common one-year scale.

As a result, we obtain charts that are well suited to get an overview of seasonal trends. For example, we can easily see the expected pattern that the `hours of sunshine` in Figure 2.11a peak around the day of the summer solstice on June 21st and reach the bottom around the day of the winter solstice on December 21st.

In Figure 2.11b, we can see that the `air temperature` in principle follows the `hours of sunshine`, but with a delay of approximately one month. This offset is well known and called *seasonal lag*. While not showing anything exciting so far, Figures 2.11a and 2.11b still helped us to detect the expected.

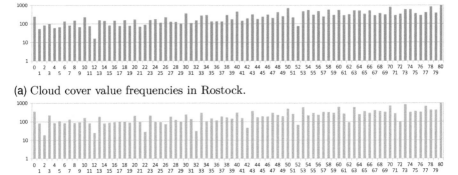

(a) Cloud cover value frequencies in Rostock.

(b) Cloud cover value frequencies in Dresden.

Figure 2.12 Histograms of the distribution of cloud cover values.

This gives credence to the validity of the data and the expressiveness of the visual representation.

But we can also discover unexpected things. Looking at Figure 2.11c, the average daily `cloud cover` (with 0/80 and 80/80 meaning clear sky and fully covered, respectively) seems almost evenly distributed over the course of the year at a first glance, with just a slight tendency of denser sky cover in the winter months. But if you look closely at the chart, you might be able to spot a horizontal line that is lighter and less covered with dots than all others.

Could it be that there are unusually few occurrences of a particular cloud cover value? As we reason about this question in more detail, we change our task from *exploring* seasonal trends in the data to *describing* the unexpected finding. Providing a description of the finding may help those who were unable to discern it on their own: The thin horizontal line is located a little above the 50/80 mark for an average `cloud cover` of 52/80.

Next, we want to find out whether our finding is by itself significant, or if we just interpret too much into the noise among the data. This implies a slight shift in our task from *describing* the unexpected finding to *explaining* it. To this end, we study the frequency distribution of cloud cover values, that is, we look at how frequently individual values appear in the data.

Again, we approach our analytic task with visual means, more concretely with *histograms* as shown in Figure 2.12. A histogram shows along its horizontal axis the individual data values whose frequencies are of interest to us. For each of the 81 possible `cloud cover` values from 0 through 80, there is a vertical bar whose height encodes how often a value occurs in the data. The histogram clearly shows that values of 52/80 and also of 12/80 are much rarer than all others. Hence, our observation was not a visual illusion, but is backed by the data.

Finally, we want to *confirm* our finding and show that it is not specific to the time series at hand. For this purpose, we look for similar patterns in the time series from a different weather station as shown in Figure 2.12b. And indeed, we can see a similar phenomenon. The pattern does not only re-occur for our initial observation at 52/80 and 12/80, but it even seems to affect the data more broadly at 2/80, 22/80, 32/80, 42/80, and so forth. At this point, we can only speculate about the reasons behind this pattern. For a satisfactory explanation, further investigations would be necessary, which suggest that visual data analysis is often an open-ended process with many twists and turns involving many analytic tasks.

With the above example, we end the discussion on influencing factors. Next, we continue with conceptual considerations regarding the fundamental processes of interactive visual data analysis.

2.3 PROCESS MODELS

In the previous sections, we have introduced criteria and influencing factors. The criteria help us to capture a notion of the quality of analysis solutions, whereas the influencing factors tell us what we need to consider in order to obtain high-quality results. What we have not talked about so far is the actual solution itself. How does it come about, how does it look like, and how is it employed? All these questions will be dealt with in this section.

In the following, we will shed light on the processes that the different actors being involved in the context of interactive visual data analysis have to deal with. In the first place, there are *designers* who are responsible for conceptualizing data analysis solutions. They need to know about the design approach of how to get from an application problem to an actual solution for that problem. Secondly, there are the *developers* who engineer and implement the solution to create a running system. Developers need a conceptual blueprint of the process of transforming data artifacts through several stages into interactive visual representations. Finally, target *users* actually analyze data. In this regard, it is of interest how utilizing data analysis solutions leads to new insights. Understanding this aspect will enable us to deploy and embed our solutions successfully into real-world problem-solving workflows.

Following the above lines of thinking, we will next study abstract models of interactive visual data analysis from a design perspective, a data-transformation perspective, and a knowledge-generation perspective.

2.3.1 Design

Designing interactive visual data analysis solutions is a challenging endeavor. The key difficulty is to come up with an appropriate ensemble of visual, interactive, and computational means that actually help users in accomplishing application-specific analysis problems.

To cope with this difficulty, it makes sense to follow a well-structured design process called the *nested model* [Mun09; Mey+15]. The model consists

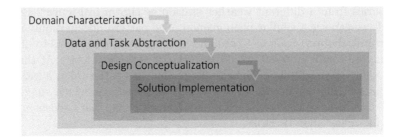

Figure 2.13 Four nested levels outline how to design interactive visual data analysis solutions. Adapted from [Mun09].

of four nested levels that describe the way from a domain problem to an actual implementation of a solution. Here, we describe a variant of the nested model that has been slightly adapted, mainly to be consistent with the terminology used in this book. The model's levels are illustrated in Figure 2.13 and are detailed below.

Domain Characterization First, the visualization designers must familiarize themselves with the application domain. This primarily concerns understanding the users and their domain-specific problems. Moreover, it is necessary to characterize the context of the data analysis as discussed in Section 2.2.3. What are the typical workflows and what tools are employed? How does the working environment look like? It is mandatory that these questions be answered in cooperation with domain experts.

Data and Task Abstraction Once the application domain is characterized, the next step is to abstract the essential data and tasks based on the notions introduced in Sections 2.2.1 and 2.2.2. This is to focus the design on conceptually relevant aspects, rather than domain-specific details. For example, instead of speaking of biomedical signaling pathways, we would say we are dealing with time-varying attributes in graphs $T \times R \to A$. The diffuse application problem of trying to *understand signaling pathways*, for example, would be translated to *explore recurring pattern of peaks visually* and *confirm them computationally*.

The abstract description of data and tasks will then inform the design conceptualization on the next level.

Design Conceptualization This level is about defining the appropriate means that constitute the overall solution. As discussed earlier, the means can be of visual, interactive, or computational nature. That said, the design conceptualization involves finding expressive visual encodings, useful interactive tools, and potent computing procedures.

Thanks to the abstract data and task description established in the preceding step, we can identify approaches that have previously proven successful, and if necessary, adapt them to the application domain. Only if no prior art exists is it necessary to develop entirely new approaches.

As we will see in Chapters 3 to 5, the design space for visual, interactive, and computational approaches is considerably large. This implies that substantial expertise is necessary in order to make appropriate design decisions that match the data, the tasks, and the context.

Solution Implementation Finally, the conceptualized design has to be brought to life. To this end, the building blocks identified in the previous level are implemented as concrete algorithms. They perform all the computations and transformations that are necessary for the data analysis. Moreover, the algorithms provide parameters that enable us to flexibly adjust the data analysis.

Following the nested model, as outlined above, has several advantages. First, the domain characterization suggests a *user-centered design* strategy where the needs of the users are central to later design decisions. Second, the abstraction of data and tasks bridges the gap between the different vocabularies of domain experts and the designer. The abstraction further allows us to categorize application-specific problems into classes of data analysis problems, which in turn simplifies the selection and the development of reusable approaches. Finally, the separation of concerns between conceptualization and implementation is beneficial because it allows the designer to concentrate on identifying or devising appropriate techniques, and the developer to focus on efficient algorithms.

As another major advantage, the nested model facilitates testing the validity of a data analysis solution. Why is this important? The problem is that the design process is prone to incorrect design decisions. Being aware of the potential threats will help us to carefully check and critically question the design choices we make. Next, we briefly describe what can compromise the validity of our design on the different levels.

- *Inaccurate domain characterization.* The application domain has not been characterized accurately. For example, the identified problems are not actual problems of the target users, or their workflows and working environment deviates from what was understood.

- *Incorrect data and task abstraction.* The abstraction fails to extract the correct descriptions of the data and the tasks. The data may contain different information than expected. The tasks are not appropriate to successfully approach the application problem.

- *Inadequate design conceptualization.* The devised techniques are not adequate. The visual encoding might not be expressive or effective. The interaction techniques could not match the environment. Computational results could be invalid due to the unfitness of analytical procedures.

- *Inefficient solution implementation.* The solution does not run efficiently. This could be due to the time complexity of the involved algorithms or their inefficient implementation. Moreover, the memory footprint could be too large for the available resources.

It is important to realize that invalid design decisions on the upper levels typically lead to a cascade of necessary changes at the lower levels. For example, a misunderstood application problem can void all later design efforts. On the other hand, an issue at the lower levels can often be remedied with moderate effort, for example, the implementation could be improved by replacing a computationally expensive exact calculation with a sufficient approximation.

In summary, the nested model describes the fundamental design process and where problems could occur along the way from the application problem to the implementation of algorithms. Next, we will switch our perspective from the overall design to the actual process of transforming data into images.

2.3.2 Data Transformation

In this section, we are interested in what actually happens to the data when becoming expressive visual representations. Certainly, we begin with raw data as input and seek images[1] as output. But how do we continue from there? What are the steps to be performed in order to transform data into images?

From a most abstract point of view, we can think of this process as a parametric transformation v that takes data and parameters as its input and generates an image as its result [JMG07]:

$$v : \mathbf{D} \times \mathbf{P} \to \mathbf{I}$$

By invoking $I = v(D, P)$, some data $D \in \mathbf{D}$ are transformed to an image $I \in \mathbf{I}$, where $P \in \mathbf{P}$ is a parameterization that controls the transformation. Let us next dig a bit deeper into what v does internally and what changes D will undergo. Models that can help us in this regard are the *visualization pipeline* [HM90] and the *data state reference model* [CR98]. They conceptually model the data-to-image transformation by defining different data stages and different types of operators.

Data Stages   From bits and bytes to images, data exist in various states. In order to abstract from particular state details, it makes sense to consider four basic data stages (note the difference between state and stage):

- data values,

- analytical abstractions,

---

[1]For brevity, we use *image* to denote any form of interactive visual representation.

- visual abstractions, and

- image data.

The data values are typically raw and unprocessed digital pieces of information. Analytical abstractions are well-structured data meaningfully enriched with derived characteristics. This includes data tables, hierarchically structured levels of detail, and higher-order abstractions such as classifications or clusters. Visual abstractions model the visual appearance of the data by means of geometric primitives and corresponding visual attributes such as fill color or stroke style. Finally, image data describe the colored pixels to be shown on the output device.

The different data stages are illustrated as boxes in Figure 2.14. The circles in the figure represent operators that do the actual data processing as described next.

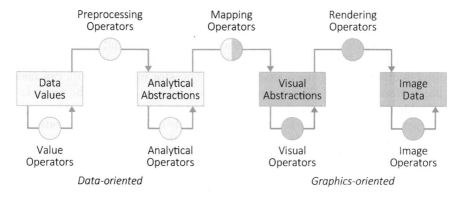

Figure 2.14 Data-oriented and graphics-oriented stages and operators. Adapted from [Chi00].

Operators An operator processes some type of input in order to produce some type of output. In the light of the above data stages, we can distinguish two categories of operators, transformation operators and stage operators.

As illustrated at the top of Figure 2.14, *transformation operators* take data from one stage and generate data of another stage. That is, the type of input an operator accepts and the type of output it produces belong to different stages. The data transformation happens in three fundamental steps: preprocessing, mapping, and rendering. These three steps form the so-called *visualization pipeline*. Along the pipeline, data-oriented operators carry over their results to graphics-oriented operators, as indicated by different colors in Figure 2.14. For example, relevant wind features are extracted from radar data (preprocessing), then color-coded onto a cartographic terrain model (mapping), and then finally displayed on a stereoscopic projector (rendering).

For *stage operators*, as illustrated at the bottom of Figure 2.14, the type of input and output are within the same stage (not necessarily of the same type). Operators that handle data values are usually responsible for data cleansing and format conversions. At the stage of analytical abstractions, further data processing can take place. For example, metrics may be calculated to determine the quality of clusters, or a hierarchical structure of different levels of detail may be established. Stage operators on visual abstractions support the adaptation of the visual representation, for example, by re-arranging visual elements or by reducing their numbers so that they fit the display. Finally, image operators may modify the graphical output. They can blur certain parts of the image to attenuate them, or enhance contrast and lightness of other parts for emphasis.

With the help of operators and data stages, we can now define the internals of our previously sketched mapping $v : \mathbf{D} \times \mathbf{P} \to \mathbf{I}$ as a network of interlinked processing steps that transform data from one state to another across several stages as illustrated in Figure 2.15. It is this network that needs to be set up when designing interactive visual analysis solutions.

As mentioned earlier, parameters are used to adjust and control the data transformation. While not being directly visible in our example, the parameters are essentially part of the operators. Each operator in the network can make parameters available. These can then be set interactively by the user via a suitable user interface, or they can be set automatically based on the output of operators of the network. Defining reasonable default parameterizations is also a task to be dealt with during the design phase.

Finally, note that the presented model is not restricted to a single input being transformed to a single output. In practice, it is not uncommon to deal with several data sources, and the data often contain so much information that multiple images are needed to represent them in an understandable way. In our example in Figure 2.15, D consists of two data sources and I comprises actually two visual representations.

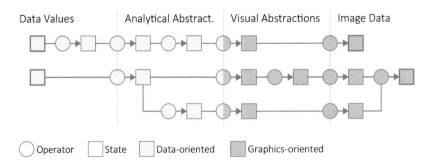

Figure 2.15 A network of operators describes the data's transformation through several stages from data values to image data.

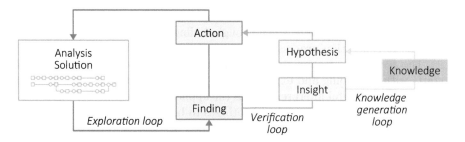

Figure 2.16   Knowledge generation model. Adapted from [Sac+14].

We have learned how data analysis solutions can be modeled as operator networks that transform data to interactive visual representations. In Chapters 3 to 5 of this book, we will learn more about concrete visual designs, interactive techniques, and computational procedures. For now, let us make a switch from the internals of the data transformation to studying how using interactive visual data analysis solutions leads to new insight and knowledge.

2.3.3   Knowledge Generation

Developing useful data analysis solutions is essentially a human-driven effort. The transformation from data to expressive images takes place in the machine. Finally, it is again the human's turn to make sense of the depicted data. What we are actually interested in is the gain in knowledge generated by interactive visual data analysis. This gain does not pop up all of a sudden, but is the outcome of an iterative process [vWij06].

Here, we cannot delve into the intricate mechanisms of the human brain, which, as already mentioned, are not yet fully understood. Instead, we consider the *knowledge-generation model of visual analytics*, which conceptually describes the interplay of the human and the analysis solution [Sac+14].

The analysis solution is shown as a single block on the left side of Figure 2.16. Despite being depicted here rather abstractly, the solution internally consists of all the interactively parameterizable data abstractions, processing steps, and visual encodings as outlined in the previous section.

In this section, we want to concentrate on the model's human part as illustrated on the right side of Figure 2.16. The human is connected to the analysis solution via three loops. These loops suggest that knowledge generation takes place at different levels of sophistication.

The first loop is the *exploration loop*, which primarily supports exploratory goals. The loop starts with observing the output generated by the analysis solution. Interesting findings made during the observation will prompt the user to take actions. Interesting findings can be trends or recurring patterns, but also the inability to detect anything useful can trigger actions. Actions can be understood abstractly as adjustments of the data transformation process in terms of configuring the network of operators or altering the operators' parameters. At some point, when enough findings have been accumulated, the knowledge-generation process will transition into the verification loop.

The *verification loop* models the confirmatory phase of the data analysis. It starts with describing and explaining the findings to constitute insight in the sense of a meaningful interpretation in the application domain. The application domain is also the source of the hypotheses to be tested. As we gain more and more insight, new hypotheses can be formed or existing ones be confirmed or rejected. This involves carrying out actions to collect sufficient evidence. For example, additional calculations can be performed to check whether they produce similar results, or alternative visual representations can be generated to see whether they support the same conclusions.

Finally, the third loop is the actual *knowledge-generation loop*. At this point, we are on the verge of turning the accumulated insight into new knowledge of the application domain. Making this final step usually requires consultation with domain experts to determine whether sufficiently strong evidence exists to trust the analysis results. If this is not the case, either additional insight must be gained from the data, or new or altered hypotheses need to be established and tested.

The three loops of exploration, verification, and knowledge generation give us but an idea of how interactive visual representations of data are transformed into something trustworthy and valuable. In practice, the reasoning process is rather spontaneous and does not follow strict deterministic rules. During the analysis, the human constantly switches between the three loops as new findings and insight lead to the formulation of new questions. For example, new knowledge may give rise to new hypotheses to be verified. In turn, the rejection of a hypothesis might initiate further rounds of exploratory activities.

With the knowledge-generation model as outlined above, we have completed our study of the processes behind interactive visual data analysis. The three perspectives we have taken, the design perspective, the data-transformation perspective, and the knowledge-generation perspective, provide us with a sound overview of the interplay of human and computer efforts when it comes to setting up, implementing, and using data analysis solutions. The variety of topics touched upon in this section suggests that experts from several fields should work together in order to make the data analysis a success, including experts from cognitive sciences, visualization, human-computer interaction, data mining, and from the problem domain itself.

2.4 SUMMARY

The development of interactive visual data analysis solutions is a challenging endeavor. It requires observing several aspects. In this section, we focused on criteria, influencing factors, and process models. Below, we briefly summarize the key messages of this chapter.

Criteria We established three fundamental criteria that interactive visual data analysis solutions should satisfy. The expressiveness criterion tells us that interactive visual representations must communicate the relevant information and must afford the necessary actions to generate knowledge. The effectiveness criterion demands that analysis solutions be aligned with the human visual and motor systems in order to be able to extract information and perform interactions effectively. Finally, the efficiency criterion suggests balancing benefits and cost of the data analysis.

Influencing Factors In order to develop solutions that meet the aforementioned criteria, we must take several influencing factors into account. First of all, the analysis is influenced by the data we are dealing with. We discussed several characteristics of data. At the level of data values, we distinguished data domains with different scales. Entire datasets are characterized by the data structure, the data space (with dimensions and attributes as dependent and independent variables, respectively), the data size, and the data scope. We also considered additional meta-data that describe the data's evolution.

Another important influencing factor is the tasks to be accomplished during the data analysis. We characterized tasks by four facets: goals, questions, targets, and means. The goal captures what the general purpose of the analysis is. The questions describe concretely what should be answered by the analysis. The targets tell us where in the data we should focus our analysis effort. The means suggest that tasks can be accomplished in various ways by visual, interactive, or computational methods, or combinations of them.

Finally, the user and the technology are important influencing factors. With regard to the users, we must observe general human factors, individual backgrounds and expertise, the application domain, and whether users work individually or collaboratively. Regarding the technology, we need to consider the available computational resources, display devices, and interaction modalities.

Process Models Three fundamental processes are at work in the context of interactive visual data analysis: the design process, the data-transformation process, and the knowledge-generation process.

The design process can be modeled along four nested levels, in which domain problems are collected, abstract data and task descriptions are derived, appropriate visual, interactive, and computational means are composed, and suitable algorithms are implemented.

The data-transformation process describes the way of the data from their raw form to interactive visual representations. Along the path, the data pass the stages of input data, analytical abstractions, visual abstractions and image data, the former ones being data-oriented, the latter ones being graphics-oriented. A network of operators is responsible for the actual data processing and data transformations.

Finally, we introduced a model that abstractly outlines the knowledge-generation process by means of three intertwined loops. These loops subsume exploratory and confirmatory analysis activities culminating in the creation of new domain knowledge.

In summary, this chapter presented the fundamental know-how that is necessary to develop interactive visual data analysis solutions. In the following chapters, we will describe in detail visual, interactive, and computational approaches that are useful for analytic purposes.

FURTHER READING

General Literature: [Mac86] • [vWij06] • [War12] • [Gua13] • [Mun14]

Visual Analysis Tasks: [Vic99] • [Sch+13a] • [Bre16] • [KK17] • [LTM18]

CHAPTER 3

Visualization Methods and Techniques

CONTENTS

V ISUAL representations of data are at the heart of this chapter. Theoretically, there are myriad ways of mapping data to visual representations. Some will lead to nice depictions of the data and might even have some aesthetic value, whereas others might be less successful.

This suggests that the visual mapping is most crucial. Different mapping strategies lead to distinct visual representations that communicate the features of the data quite differently [Han09]:

"The representation effect: Human performance
varies enormously (10–100:1) with different representations."

Hanrahan, 2009

The *representation effect* is illustrated in Figure 3.1. It shows three plots, all visualize the same data, a daily time series of the number of people diagnosed with an influenza-related illnesses. Yet, the three visual representations were generated using different visualization techniques and different parameterizations. Figure 3.1a shows a linear plot. We can clearly identify some peaks in time and periods of moderate and low number of cases. It seems there is no clear trend over time.

But are there any cyclic patterns? To answer this question, a different visualization technique can be employed: a spiral plot. The spiral plot in Figure 3.1b uses a cycle length of 32 days. From this figure, no cyclic patterns can be discerned. So let us try a different parameterization. In Figure 3.1c, the cycle length has been set to 28 days, which amount to exactly four weeks. As the weekdays are aligned now, we can see a clear cyclic pattern: At the beginning of a week more people are diagnosed and hardly any diagnoses are reported on weekends. This pattern is impossible to discern from the linear plot. On the other hand, the exact peak times are not as easy to identify in the spiral plot.

By this example with quite simple data and rather basic visualization techniques, we have seen how important the design of visual representations is. Depending on what features of the data are to be communicated to answer which questions (e.g., Where are the peaks? Or, do cyclic patterns exist?), different options may be appropriate.

In this chapter, we are interested in visual representations that fulfill the criteria of expressiveness, effectiveness, and efficiency as introduced in the previous chapter. To this end, it is important to learn basic visual building blocks and understand how to compile them into visual data representations. The two key concerns when designing a visual representation is to decide how to encode data visually and how to present the data meaningfully to the user. The fundamental ideas behind visual encoding and presentation will be described in Section 3.1.

Once a suitable design has been devised, it can be implemented as a visualization technique. Visualization techniques typically make some assumptions with respect to the data for which they are applicable. Parameters may be available to fine-tune the visual representation within reasonable bounds.

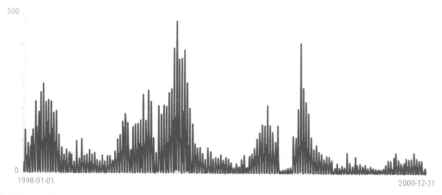

(a) Line plot.

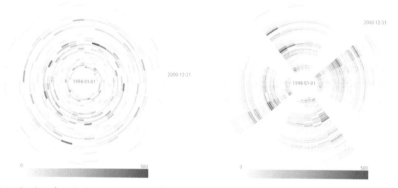

(b) Spiral plot (cycle length 32 days).   (c) Spiral plot (cycle length 28 days).

Figure 3.1   Illustration of the effect of different visual representations.

Because there are different classes of data, there are different visualization techniques, of which this chapter offers a data-oriented overview. First, we discuss basic techniques for visualizing multivariate data in Section 3.2. Then our view will be extended in Sections 3.3 and 3.4 to techniques for temporal data and geo-spatial data, which require the communication of a temporal and a spatial frame of reference, respectively. Graph data define relations among data elements, which calls for dedicated visualization techniques as we will see in Section 3.5.

In short, we broaden our view step-by-step from basic visual encodings to the visualization of multivariate data attributes A, to techniques for data with temporal context T and spatial context S, to relations among the data R. This also includes different combinations of the individual aspects.

3.1 VISUAL ENCODING AND PRESENTATION

In 1967, based on an extensive analysis of visual representations in cartography, Jacques Bertin introduced the idea of data graphics expressed by marks and visual variables [Ber67; Ber83]. *Marks* serve as the carriers of information. They are distinguished by their dimensionality. There are 0D points, 1D lines, 2D areas, and 3D bodies. Marks are the basic building blocks of visual representations.

The actual information is conveyed via *visual variables* such as position, shape, or hue, which control the marks' visual appearance. In other words, visualizing data means creating graphical primitives and specifying their visual appearance according to the underlying data values. This concept of data graphics is still the fundamental basis of visualization.

3.1.1 Encoding Data Values

Let us discuss the visual encoding with visual variables in more detail. A visual variable can be understood as a graphical property that can be varied within a reasonable range. Different variations of a visual variable can be perceived as different. For example, we can discern position, shape, or hue of a mark. Originally introduced by Bertin, the concept of visual variables has later been refined and extended in several ways [Mac86; Mac95]. Figure 3.2 illustrates a selection of visual variables commonly applied today.

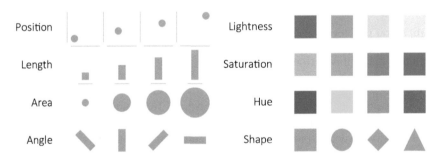

Figure 3.2 Visual variables.

With the help of visual variables, we can define the visual encoding, that is, how the data are represented visually. In this regard, two questions must be addressed:

- What to map?

- How to map?

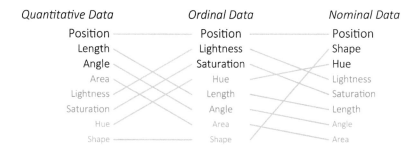

Figure 3.3 Effectiveness ranking of visual variables. Adapted from [MW14].

What to Map?

First, we must decide which data variables are to be mapped onto which visual variables. For answering this question, we need to consider the analysis task at hand, in particular the task's target as discussed in Section 2.2.2. It tells us which data variables are deemed relevant and hence should be mapped onto the most effective visual variables. The question that remains is which visual variables are the most effective? As we will see next, the answer depends on both, human perception and the data domains introduced in Section 2.2.1.

With the help of perceptual studies, researchers found that visual variables are differently effective for nominal, ordinal, and quantitative data [CM84; HB10]. Visual variables that allow us to make precise distinctions are particularly suited to encode nominal data. Visual variables that facilitate a visual ordering are well suited for ordinal data. For quantitative data, we need visual variables that support the estimation of proportions or differences.

An effectiveness ranking of visual variables is illustrated in Figure 3.3. Interestingly, encoding to position is effective for all data domains. On the other hand, a visual variable's rank may change drastically across the different data domains, as it is for instance the case for hue and shape, which are best applied for nominal data but are less or not effective for ordinal and quantitative data.

How to Map?

Once we have decided what to map, we must think about how the data are to be mapped onto a visual variable. As already indicated in Chapter 2, the task and the data at hand influence the mapping strategy. Figure 2.8 back on page 29 already illustrated the utility of different color mappings for different tasks. In the following, we will briefly look at further ways of tuning the visual mapping to the task and the data. For the sake of brevity, we will concentrate on the use of colors.

Identifying and locating data values on the display are fundamental analysis tasks. In order to support identification tasks, *perceptual* color maps should be

Figure 3.4   Color maps for identifying and locating values and classes.

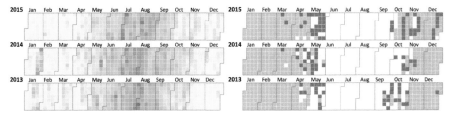

(a) Color coding for identification tasks.   (b) Color coding for location tasks.

Figure 3.5   Applying the color maps from Figure 3.4 to temperature data. Adapted from `bl.ocks.org/mbostock/4063318`.

used. Such color maps guarantee that perceived color differences are proportional to the underlying differences in the data [BRT95]. Ideally, there should be a clearly perceptible color for each value in the data. Certainly, the degree to which this ideal can be reached is limited by the output device and human perception.

For location tasks, the goal is to determine where in the visual representation a certain value or subrange of interest is located. This is supported by color maps that let the relevant data stand out. Again, the characteristics of human perception need to be considered. Particularly helpful are visual encodings that are pre-attentive [HE12]. For example, the relevant data are encoded with a signal color, whereas less-relevant parts of the data are visualized with dimmed colors.

Another important factor relates to the question whether the analysis targets individual data values or classes of values. Continuous color maps are a sensible option for individual values. Segmented color maps are suitable for classes of values. Figure 3.4 provides a schematic illustration with color maps for identifying and locating individual values and classes of them. In Figure 3.5, we applied two of these color maps for the visualization of three years of daily temperatures. Figure 3.5a uses the continuous color map to support the identification of data values, whereas Figure 3.5b uses a segmented color map to support the locating of a particular class of values, days with temperatures between ten and twelve degrees Celsius in our case. What we can see from these figures is how substantially different the data look like when using different color maps. What we can't see is how the mapping works internally. This will be explained next.

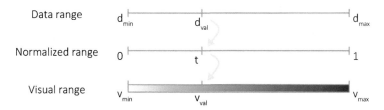

Figure 3.6 Basic mapping of a data variable onto a visual variable.

Typically, the entire data range from the minimum value to the maximum value is mapped onto the entire spectrum of a visual variable. A linear mapping as expressed by the following formulas is common to assign a visual value v_{val} (a color in our case) to a data value d_{val}. In a first step, it is practical to calculate a normalized data value $t \in [0, 1]$:

$$t = \frac{d_{val} - d_{min}}{d_{max} - d_{min}}$$

This normalized t is then used in a convex combination to compute the actual visual representative v_{val} from the visual range $[v_{min}, v_{max}]$:

$$v_{val} = (1 - t) \cdot v_{min} + t \cdot v_{max}$$

Figure 3.6 illustrates the basic mapping of a quantitative variable onto a continuous color map. Note that for qualitative data, it is necessary to first map categorical or ordinal values to numbers, for which dedicated methods exist [Ros+04]. The distinction between quantitative and qualitative data is not the only factor that influences the mapping strategy. In the following, we will illustrate this with additional examples.

As a first example of a data-dependent adaptation, let us consider the scales shown in Figure 3.7a. Scales play an important role when it comes to interpreting visual representations, that is, when the user performs the inverse mapping of color to data. Plainly using the data's actual minimum and maximum often leads to situations where the interpretation is unnecessarily complicated. As can be seen in our figure, expanding the mapping range $[d_{min}, d_{max}]$ to multiples of two or powers of ten can significantly ease the interpretation [JTS08]. Several algorithms exist to create similar results [TLH10].

In addition to the data's minimum and maximum, also the magnitude spanned by the value range influences the visual mapping. The basic mapping introduced before works well in many cases. However, for large value ranges that cover several magnitudes and include small as well as large values, a linear mapping could be ineffective. In such cases, it makes sense to use a non-linear mapping. Figure 3.7b compares a linear-scale mapping with a log-scale mapping. For the logarithmic variant, the calculation of t changes as follows:

$$t = \frac{\log(d_{val} - d_{min})}{\log(d_{max} - d_{min})}$$

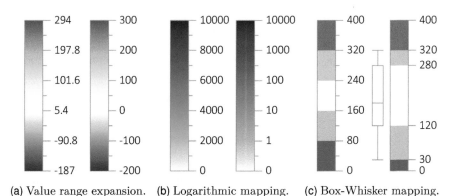

(a) Value range expansion.   (b) Logarithmic mapping.   (c) Box-Whisker mapping.

Figure 3.7   Enhanced data-dependent visual mapping.

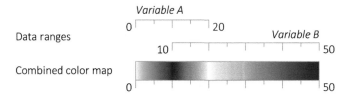

Figure 3.8   Combined color map for comparing two data variables.

In situations where the analysis is focused on semantically meaningful ranges of values, it can even make sense to go beyond linear or logarithmic mapping. Figure 3.7c shows a basic mapping in contrast to a mapping that has been adapted based on the data's underlying distribution. In the example, the distribution is approximated via quartiles and inter-quartile ranges in the data as indicated by the box plot [Tuk77]. As a result, we obtain a visual mapping that emphasizes unusually low and high data values and outliers with shades of blue and red, whereas common values are represented by neutral white.

As a last example, let us consider the visual mapping when the task is to compare data, for example, two data variables or data given at two time steps. The challenge for the visual mapping lies in balancing two conflicting needs. On the one hand, the individual sets being compared should be distinguishable and values should be identifiable. This calls for a separate visual encoding per set being compared. On the other hand, the sets must be comparable. However, this is usually not the case when using separate visual encodings. Instead, a common encoding must be employed for all data, so that the same data value, no matter with which data variable or time step it is associated, is always visualized with the same visual stimulus. However, this could lead to local data variations being washed out in the visual representation.

One way to address these conflicting needs is to combine multiple color maps into a dedicated comparison map [TFS08b]. Figure 3.8 illustrates this for comparing two data variables A and B. Variable A has lower values and is mapped using variations of green, whereas variable B has larger values represented by variations of blue. Shades of gray indicate the range of values that both variables have in common. With the help of the combined color map, comparison is possible, while individual values are still distinguishable.

In summary, we see that the decision on how to map data values onto visual variables must be done with care. What has been said here about color coding applies analogously to other visual variables. More details about color coding in general and specifically for different tasks and data characteristics can be found in the further readings at the end of this chapter. In the following, we continue with advanced mapping strategies in which multiple visual variables are employed.

Using Multiple Visual Variables

A 1-to-1 visual mapping is the standard approach to visual encoding: A single data variable is mapped onto a single visual variable. Yet, it is also possible to utilize n visual variables simultaneously. There are two variants of this approach. A 1-to-n visual mapping takes advantage of the combined power of n visual variables for the encoding a single data variable. When multiple data variables are to be visualized, an n-to-n visual mapping has to be designed. Note that n-to-1 visual mapping makes no sense, because it would no longer be possible to unambiguously interpret a visual stimulus.

1-to-n Visual Mapping   In general, we can use several visual variables to encode a data value. For example, we can set a mark's position, size, and color according to one and the same data value. This typically makes the value easier to recognize.

A particularly clever approach of combining $n = 2$ visual variables is two-tone coloring [Sai+05]. The goal of two-tone coloring is to achieve precise value recognition while keeping space demands low. Figure 3.9 illustrates the problem. When visualizing data as a plot, values can be read precisely. However, a certain amount of vertical display space is needed in order to make data values easily recognizable. Less display space is needed when the data are visualized via color coding. Yet, the data values are more difficult to read. With a continuous color map, it takes some time to figure out which value is associated with a color. With the segmented color map, only discrete classes of values can be identified. Let's look at how two-tone coloring deals with these issues.

The basic idea is to encode a data value with two colors and combine this with a length encoding. As shown in the enlarged view on the right-hand side of Figure 3.9, a data value is visualized by two adjacent colors from a segmented color map, orange and red for the particular data value in our example. The colors tell us at a glance where in the value range the specific data value is

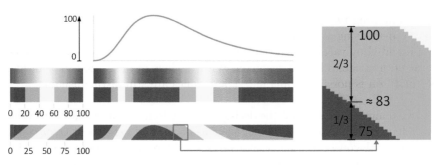

Figure 3.9   Two-tone coloring explained. Adapted from [JTS08].

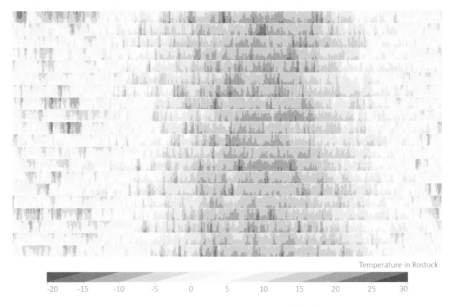

Figure 3.10   Two-tone visualization of 20 years of daily temperatures.

located. It is between 75 and 100. The precise value is encoded by varying the proportion of the colors, that is, how many pixels are colored with the first color and how many with the second color. In the example, 1/3 of the pixels are red and 2/3 are orange. This means the encoded value is one-third between 75 and 100, which is approximately 83.

Figure 3.10 illustrates that the two-tone approach has several advantages. First, thanks to the combined use of color and length as visual variables, less display space is necessary, which allows us to overview more data. Second, colors make it easy to locate data of interest. Third, it is even possible to rather accurately identify individual values. In sum, a quite efficient visual encoding.

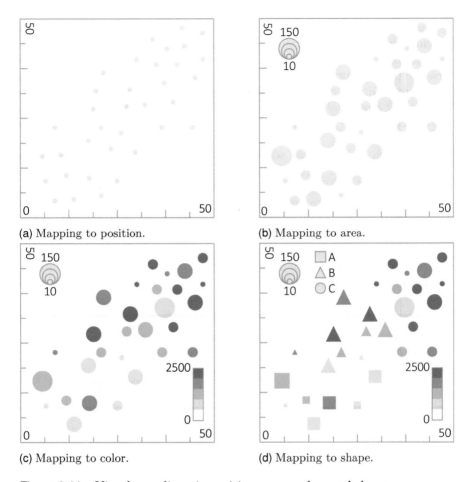

Figure 3.11 Visual encoding via position, area, color, and shape.

n-to-n Visual Mapping   We already mentioned that it is often necessary to visualize more than one data variable. In such cases, the visual encoding can utilize several visual variables simultaneously. This shall be illustrated by the example of a scatter plot.

A scatter plot consists of two orthogonally aligned axes that represent the value ranges of two data variables. Dots are placed in the space spanned by the axes in order to visualize the data elements. Conceptually, this corresponds to a mapping of data to position. A first data variable is mapped with respect to the horizontal x-axis, and a second variable with respect to the vertical y-axis. Figure 3.11a shows a basic example.

In order to encode additional data variables, we can now employ additional visual variables. For example, we can vary the area of the dots to encode a quantitative data variable as in Figure 3.11b. With larger dots, it is also possible to use colors to visualize an ordinal data variable, which is illustrated in Figure 3.11c. We can even think of replacing the dots with different shapes to visualize a nominal data variable. In sum, Figure 3.11d now represents five data variables: two are mapped to position, one to area, one to color, and a fifth to shape.

With each variable being added to the visual mapping, the richness of the visual representation is increased. Theoretically, we could add yet another visual variable, for example, by texturing the shapes. However, from a practical point of view, there are limits. While a rich visual mapping opens up the possibility to make a wider range of analytic discoveries, the downside is that the mental effort required to digest the visual representation increases as well. Therefore, it is really important to balance the visual mapping according to the task and the data. When designing a visualization, the following quote can be useful to keep in mind [dSai39]:

"In anything at all, perfection is finally attained not when there is no longer anything to add, but when there is no longer anything to take away [..]"

de Saint-Exupéry, 1939

In fact, the visual encoding is one of the most crucial steps when designing visual data analysis solutions, because it has considerable impact on expressiveness, effectiveness, and efficiency. However, a good visual encoding is only half the story. Of equal importance is an appropriate presentation of the encoding, as we will see in the next section.

3.1.2 Presentation

The visual encoding produces marks whose graphical appearance meaningfully reflects the data. The marks are compiled into views to be displayed to the user. The question now is how to present the marks and views in such a way that the visualized data are easy to understand. In the following, we will discuss key questions that need to be taken into account when designing the presentation. Should the marks be presented in 2D or 3D? When dealing with large numbers of data elements, should we display the marks for all of them or should we concentrate on subsets of interest? And, how can we present multiple views each dedicated to showing a particular aspect, perspective, or facet of the data?

Presenting Marks in 2D or 3D

The first design decision is whether to place the information-bearing marks in a two-dimensional (2D) or a three-dimensional (3D) presentation space. A 2D presentation space has two independent axes: the horizontal x-axis and the vertical y-axis. A 3D presentation space has an additional third z-axis.

Both 2D and 3D representations have different pros and cons. Two-dimensional visualizations are arguably more abstract and easier to explore. On the other hand, human perception is naturally tuned to the three-dimensional world around us. Moreover, the third dimension can also serve as a carrier of additional information. However, the third dimension also introduces difficulties that are less problematic or do not even occur in 2D, such as occlusions and perspective distortions. There is no definite answer whether to use 2D or 3D. The decision has to be made carefully on a case-by-case basis, taking into account several factors, including the number of marks to be arranged, the analysis task, and the available display technology. Figure 3.12 shows an example 3D terrain visualization with a 2D overview map. We will return to the implications of 2D and 3D representations later in Section 3.4 when we talk about geographic visualization in more detail.

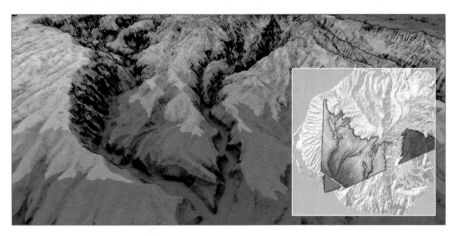

Figure 3.12 Terrain visualization with overview+detail. Courtesy of Martin Röhlig.

Presenting All Data or Data of Interest

In order to facilitate a comprehensive understanding of the data being studied, two essential communicative goals must be supported by the presentation:

1. Convey an overall picture

2. Provide details where necessary

The overall picture allows users to observe global patterns and general properties of the data. The details enable users to study local particularities of the data. For data of small or moderate size, overview and details for all data can be presented in a single view. However, with larger data, this becomes more and more difficult. For large data, conveying an overall picture usually requires neglecting details and giving preference to the presentation of the data as a whole. When details are to be presented, it is usually necessary to sacrifice completeness and concentrate on selected data.

A basic strategy to address this problem is to distinguish between data of interest and less relevant data. The idea is that marks encoding the data of interest will be presented in full detail, while the others are reduced or simplified. Two approaches support this strategy: *overview+detail* and *focus+context* [CKB08].

Overview+Detail presents an abstracted overview of all data in one view and selected details in separate views. There can be a small overview being superimposed on a large detail view, as commonly seen for geographic maps. Alternatively, a large overview can be superimposed by multiple detailed views. In Figure 3.12, the main view shows a detailed representation of the 3D terrain for a selected region, while a smaller window provides an abstracted 2D overview and highlights the region whose details are currently visible.

Focus+Context integrates focused details within a global context in a single view. The integration can be achieved in different ways [Hau06]. A classic approach is to use distortions to magnify the focus, scale down the context, and create a smooth transition in between. Fish-eye distortion is a prominent example [LA94]. Figure 3.13 illustrates a fish-eye distortion being applied to the rows of a table-based visualization of the Iris flower dataset. As can be seen, the distortion makes individual rows in the focus region easier to read and even allows text labels to be displayed.

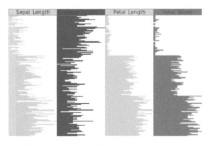
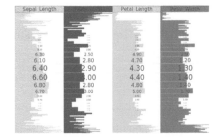

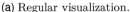

(a) Regular visualization.   (b) Focus+context distortion of rows.

Figure 3.13   Illustration of focus+context for a table-based visualization of the Iris flower dataset. Focused rows are magnified to accommodate labels.

Both overview+detail and focus+context are viable approaches to presenting the overall picture and details of larger data. For overview+detail, the user has to establish links between the separate views mentally. The integrative approach of focus+context relieves the user of this task. However, interpreting a densely packed focus+context presentation might require some training, especially when distortions are involved.

Presenting Multiple Views in Space or Time

When the data to be analyzed become more complex, it is no longer feasible to indiscriminately present each and every aspect of the data in a single view. When we reach this point, it makes sense to create several dedicated visual representations, each focused on communicating a particular aspect or facet of the data. The question is how several such views can be presented to the user in order to convey a comprehensive picture?

There are two basic answers to this question: The views are presented either in parallel side-by-side or sequentially one after the other. Conceptually, this corresponds to an arrangement of views in space or a sequencing of views in time. What this means concretely will be described next.

Arranging Views in Space   Showing several faceted visual representations in parallel side-by-side is commonly referred to as *multiple views* visualization [WWK00]. Later, the term *coordinated* has been prepended to indicate that the views are not used in isolation but in concert [Rob07]. In particular, coordination means that interactions performed in one view are automatically propagated to all other views.

An example of three different views for visualizing the same multivariate data is shown in Figure 3.14. Details about the depicted visualization techniques will be given very soon in Section 3.2. How multiple views can be employed for graph visualization will be explained at the end of this chapter in Section 3.5.3.

For the time being, we are interested in the spatial arrangement of views. There are two extremal positions when it comes to defining the spatial arrangement. On the one hand, one could use a fixed and provably efficient arrangement designed by an expert. On the other hand, users could be provided with the full flexibility to arrange views arbitrarily. Both extremes have their advantages and disadvantages, and both are actually applied in practice.

An interesting intermediate alternative is to allow flexibility only within certain limits. For example, it can make sense to prohibit partial overlap of views. In other words, a view should be either visible or not. This can be achieved by partitioning of the available screen space into regions, each of which may contain one or more views. While views are allowed to be flexibly resized and moved from one region to another, the overall arrangement is constrained to remain a partition, which is free of partial overlap.

Arranging multiple views on a single display is usually a task that can be accomplished with little or moderate effort. However, when views have to be

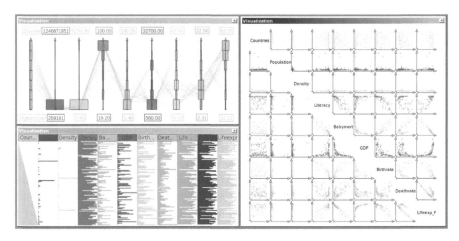

Figure 3.14   Multiple coordinated views for analyzing multivariate data.

arranged on multiple displays, the costs for defining a suitable spatial layout increase significantly. In such scenarios, users should be supported by means of algorithms that automatically distribute and lay out views on the available displays. Later in Chapter 6, we will describe such an automatic arrangement approach in detail.

Sequencing Views in Time   The alternative to arranging views in space is to sequence them in time. That is, instead of showing multiple views simultaneously, they are presented step-by-step, one after the other. Depending on how many views are shown to the user per time unit, a sequencing of views may generate the impression of a slide show or an animation.

The advantage of sequencing views in time is that each view can fully utilize the display space. There is no need to divide the space among views. Obviously, sequencing views in time is particularly suited to convey temporal characteristics of data. It can also be helpful to take the user on a journey from one data facet to another.

However, presenting views in quick succession to the user also has some limitations. For example, it could be difficult to make sense of all the information provided during a sequence of views. Especially when sequences take a long time, users may be unable to follow and could drown in an indigestible flood of visual representations. Therefore, it is mandatory to provide interactive controls to pause, slow down, reverse, and advance the presentation.

In summary, this section dealt with the basics of visual encoding. We discussed how data can be mapped to visual variables of marks, and how the marks can be presented to the user. With the general aspects of visualization being clear now, we will next study how specific techniques utilize visual variables and arrange visual marks for visualizing different kinds of data. We will start with techniques for multivariate data, that is, data with several data attributes.

3.2   MULTIVARIATE DATA VISUALIZATION

We have already seen several visualization examples in this chapter that are *univariate* in the sense that they show the data values of only one dependent data variable, for instance, the line plot and the spiral plot in Figure 3.1 or the temperature representation in Figure 3.10. However, inspecting only one dependent variable is often not enough. In many application contexts, it is necessary to discern and understand the interrelations and dependencies between multiple or even all of the dependent variables. To this end, we need *multivariate* visualization techniques.

Multivariate data visualization concentrates on the depiction of the dependent variables of a dataset, that is, the attributes A. In the following, we will introduce fundamental classes of multivariate visualization techniques. Each class is characterized by its own fundamental visualization strategy. We will consider:

- table-based visualizations,

- combined bivariate visualizations,

- polyline-based visualizations,

- glyph-based visualizations,

- pixel-based visualizations, and

- nested visualizations.

Independent variables, including time T and space S, and relationships R between data elements will be taken into account in later sections of this chapter.

3.2.1   Table-based Visualization

As detailed in Section 2.2.1, multivariate data are generally modeled as data tables where columns accommodate attributes, and rows represent data tuples. It is common to represent such data using tabular spreadsheets, which show data values as text in the spreadsheet cells. As such, spreadsheets are well-suited for reading and editing data values precisely. However, understanding the multivariate relations of the data is hardly possible. Moreover, the textual representation of data values requires considerable display space per cell, which limits the number of data tuples that can be shown.

The simple, yet very effective idea of *table-based visualization* is to retain the tabular layout of spreadsheets, but to replace the textual representation of

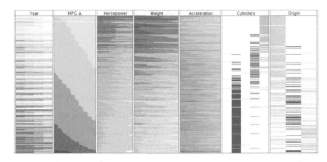

Figure 3.15 Two-tone colored table-based visualization of the Cars dataset.

Figure 3.16 Table Lens with textual labels for focused data tuples.

data values by a visual representation. A visual representation will not only make the interpretation of the data much easier, it will also require less display space.

Depending on the data domain of an attribute, different visual encodings are practical. A universally applied strategy is to color-code the table cells. An alternative for quantitative data is to embed bars into the table cells and to vary the bars' length depending on the data values. It is also possible to combine length and color analogous to the two-tone approach introduced in Figure 3.9 back on page 60. An example of a table-based visualization with two-tone coloring is given in Figure 3.15. The image shows seven properties of about 400 cars in the Cars dataset sorted by miles per gallon (MPG).

As we can see, table-based visualization can provide us with a general overview of the distribution of multivariate data. On the other hand, while we gain an overview, we lose details: The exact data values are no longer available as textual labels. The *Table Lens* is a classic focus+context technique to cope with this problem [RC94]. For a focused subset of the data, the Table Lens shows larger rows and inserts exact values in textual form. The remainder of the table, that is, the context remains unchanged. Figure 3.16 shows an example visualization of demographic data of 111 countries. The height of table rows is varied by means of a fish-eye transformation function so as to create a smooth transition between focus and context.

A table-based visualization is also compatible with the common operations available with spreadsheets. One can reorder the table columns and sort the rows according to attribute values. While this is quite helpful for studying individual attributes, finding multivariate relationships is still not an easy task. The data must be sorted successively on a column-by-column basis, and the obtained results have to be integrated mentally by the user. To reduce the required effort, it makes sense to incorporate methods that sort rows based on multivariate similarities. One such method will be presented in Section 5.4.2 of Chapter 5, which is dedicated to automatic analysis support.

3.2.2 Combined Bivariate Visualization

An alternative to table-based approaches is to visualize multivariate data by combining several bivariate displays. To this end, individual bivariate displays are created for all pairs of attributes $(a_i, a_j) : a_i, a_j \in A, i \neq j$. The bivariate displays are then combined to facilitate an overview of the entire data. The combination can be a spatial arrangement or a temporal sequence of views.

Spatial Arrangement Among the earliest and most widely used combinations of bivariate displays is the *scatter plot matrix* [Cle93]. For m attributes, the matrix is composed of $m^2 - m$ individual scatter plots, as illustrated in Figure 3.17. There are $m - 1$ plots per row and per column. Each scatter plot consists of two orthogonal axes that represent two attributes, and dots visualize the data tuples with respect to the two data attributes. The scatter plot at (i, j) visualizes the attributes a_i and a_j. The scatter plot at (j, i) shows the same data but with the axes swapped. There is usually no scatter plot at (i, i), because it hardly makes sense to plot attribute a_i against itself. Instead, the space is often used to show the value distribution of a_i or simply just a label.

The rows and columns of a scatter plot matrix can be re-arranged to support different tasks. For example, it can make sense to bring relevant attributes to the top left or to show correlated attributes next to each other.

Temporal Sequencing Instead of using a spatial arrangement as a matrix, it is also possible to create an animated sequence of scatter plots, which is known as the *Grand Tour* [Asi85]. The idea is to show the individual scatter plots one after the other. The sequence of plots is carefully chosen so that the most important multivariate relationships are revealed during the animation.

When working with combined bivariate visualizations, it is important to realize that each bivariate display communicates the data values only with respect to two attributes. Hence, finding an outlier or a correlation in either bivariate display does not tell us whether the finding is indeed a multivariate

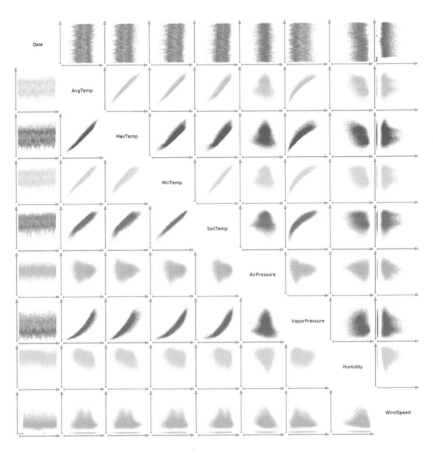

Figure 3.17 9×9 scatter plot matrix of meteorological data. Color is used to ease the recognition of data variables.

feature. In order to help users draw the right conclusions, additional mechanisms need to be integrated.

One such mechanism is interactive *brushing & linking*. It enables the user to mark dots in a bivariate display, and in this way to select data tuples. All the dots that represent data values of the selected tuples are then automatically highlighted in all bivariate displays. So, if the user marks an outlier in one display and sees that the subsequently highlighted dots are outliers in the other displays as well, then it can be assumed that indeed a multivariate outlier has been found. More details about data selection and visual highlighting are provided in Section 4.4 of Chapter 4.

3.2.3 Polyline-based Visualization

A difficulty with combined bivariate visualizations is that the connection between the individual displays has to be established by the observer mentally. That is, as the eyes move from one bivariate display to the next, the observer has to keep track of the visited dots in order to form a complete understanding of data tuples.

Visualization techniques based on polylines aim to tackle this difficulty. The basic strategy is to create m axes, one for each attribute, and n polylines, one for each data tuple. The polyline of an m-variate data tuple is constructed as follows. For each attribute value of the data tuple, a position is computed at the corresponding attribute axis. The m positions that we obtain are then connected to form the polyline that represents the entire tuple.

The question that remains is how the axes are arranged on the display. *Parallel coordinates* plots are known for their parallel arrangement of axes [Ins09]. For *radar charts*, the axes are not parallel, but emanate from a central point. Figure 3.18 illustrates both the parallel and star-shaped arrangement of axes for the same multivariate data. Axes can also be arranged in 3D, and in Section 3.3, we will see that dedicated arrangements exist for temporal data.

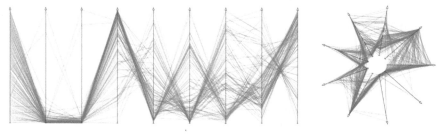

(a) Parallel coordinates plot.   (b) Radar chart.

Figure 3.18   Visualization with polylines across parallel and star-shaped axes.

It is worth mentioning that the order in which polylines connect the axes is important, because patterns in the data can be best interpreted between neighboring axes. Let us look at some of the visual patterns that are created by polyline-style visual representations. Consider the pairs of parallel axes in Figure 3.19. The visualization shows several distinct patterns synthetically created for the purpose of illustration. The first pair of axes shows a positive correlation. How a negative correlation looks like can be seen in the second pair of axes. The third and fourth pair show a peak and a valley, respectively. Finally, there are a few local outliers in the right-most pair of axes. It is interesting to think about how the same data would look like when being visualized as individual scatter plots. If you want to check if what you imagined is correct, Figure 3.20 provides the answer.

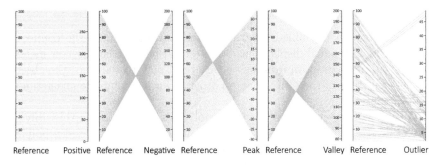

Figure 3.19 Visual patterns between pairs of parallel axes.

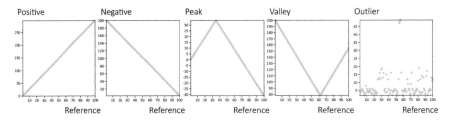

Figure 3.20 The same data as in Figure 3.19 visualized as scatter plots.

Axes-based Visualization

When thinking more deeply about scatter plots, scatter plot matrices, and parallel coordinates plots it becomes clear that they all follow the same underlying strategy: They are all based on axes with respect to which data points are projected [CvW11]. Many other types of diagrammatic representations follow this strategy [Har96]. As such, *axes-based visualization* is a universal concept that is applicable to a wide range of multivariate analysis questions.

A difficulty though is that tuples with the same data values are projected to the same positions. This leads to over-plotting. The problem is that we typically cannot tell how many tuples are represented by a line segment (or dot or pixel in general). This problem can be remedied in different ways. A simple option is to use transparency. Another option is to incorporate additional visual cues to visualize the frequency of data values. Histograms are quite useful in this regard. They can be easily attached to the axes of scatter plots or parallel coordinates plots. Figure 3.21 shows an example. Later in Chapter 5, we will return to the problem of reducing over-plotting and clutter in visual representations by means of additional automatic computations.

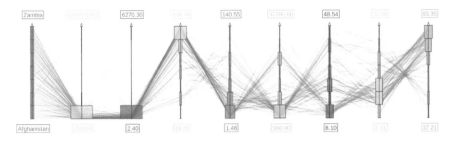

Figure 3.21   Parallel coordinates with histograms showing demographic data.

3.2.4   Glyph-based Visualization

Axes-based visualization, as discussed before, emphasizes relationships between two attributes. In contrast, glyph-based visualization aims to emphasize individual data elements. To this end, data elements are visualized as small self-contained graphical objects called *glyphs*. Each glyph visually encodes a multivariate data element, more precisely, it encodes the attribute values associated with a data element. Setting up a glyph-based visualization requires three steps:

1. Glyph design

2. Attribute-to-glyph mapping

3. Glyph placement

Glyph Design   The design of glyphs is an intricate process [Bor+13]. The particular challenge is to encode several data attributes, although there is only very little display space available per glyph.

Several questions need to be answered: How many attributes should a glyph represent? How many individual values should be discernible per attribute? Which visual variables should be used? How should the visual encodings be intertwined and the overall glyph be compiled?

When answering these questions, it must be ensured that a glyph can be easily perceived as a whole, and that multiple glyphs can be easily separated from each other. Only then can a glyph-based visualization be interpreted as individual data elements to be studied as such and compared to others.

Figure 3.22 illustrates a couple of classic glyph designs. The *autoglyph* in Figure 3.22a consists of a grid of color-coded cells, one cell per attribute. *Stick figures* as in Figure 3.22b vary the length, thickness, color, and angle of limbs in order to visualize data. The *Chernoff faces* in Figure 3.22c encode data by varying the features of a human face, such as the expression of the mouth or the eyes or the shape and size of the nose.

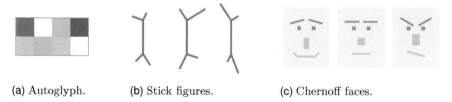

(a) Autoglyph.   (b) Stick figures.   (c) Chernoff faces.

Figure 3.22   Examples of classic glyphs for visualization.

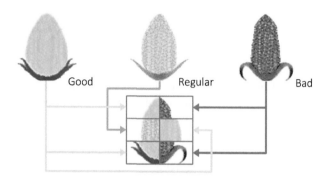

Figure 3.23   Corn glyph for representing six ordinal data values.

Attribute-to-Glyph Mapping   The second step is to carefully decide which data attributes should be visualized with which visual encoding offered by a glyph. We already discussed this mapping problem in general in Section 3.1.1. For the case of glyph-based visualization, there is one aspect that needs to be considered in particular: Usually, glyphs can encode only a few distinct values due to their limited size. As a consequence, it can be necessary to reduce the attributes' value ranges. For example, a continuous attribute may need to be discretized to only a few values before it can be mapped to a glyph.

What this means shall be illustrated by a glyph for visualizing factors influencing maize harvest. The design is based on three pictorial corncobs: a particularly well-developed exemplar, a regular cob, and a withered cob as shown in Figure 3.23. These three depictions represent three ordinal values: good, regular, bad. By cutting the corn glyph into segments, it is possible to encode several data values. However, any attribute that we would like to map to a segment of the corn glyph must first be made compatible with the ordinal scale that the glyph can display. This can be done in a classification step where numeric data values are assigned to classes.

Glyph Placement   Finally, the placement of glyphs is important. Placing glyphs on grids is useful in general. Figure 3.24a provides an example with data about the resistance of bacteria against eight antibiotics. The binary data (resistent, non-resistent) are represented as a grid of autoglyphs, where black

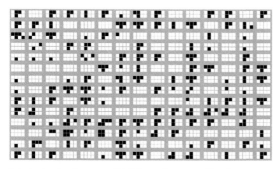

(a) Autoglyphs as grid (aka. shape coding). Reprinted from [SM00].

(b) Corn glyphs on a map. Reprinted from [NSS05].

Figure 3.24 Examples of different placements of glyphs.

means resistant and white means non-resistant. Such representations provide a nice overview of the data, which makes it possible to spot patterns, as for example, the glyphs showing no resistance at all (all cells are white) or glyphs showing a T-shape.

Glyphs can also be placed more flexibly according to data criteria. For example, the glyph placement can be computed based on a semantic grouping of data elements [War02]. If spatial dependencies exist in the data, glyphs are typically arranged so as to communicate these dependencies. Figure 3.24b shows an example with corn glyphs placed on a map [NSS05]. The use of glyphs for data with spatial dependencies will be discussed further in Section 3.4

3.2.5 Pixel-based Visualization

In contrast to glyph-based visualization, pixel-based visualization appears to be a minimalistic approach. The key idea is as simple as to create a representation where each pixel visualizes exactly one data value by its pixel color [KK94]. Yet, what appears to be minimalistic is quite the opposite. Pixel-based representations are very compact, which makes it possible to display millions of data values.

Three design questions need to be answered for pixel-based multivariate data visualization:

- How should individual data values be mapped to color?

- Where should the pixels for multiple attributes be located?

- How should the individual pixels be arranged?

The first question is particularly important. A pixel is the smallest possible graphical primitive on a computer display. In fact, a pixel is so small that identifying its color can be difficult. Moreover, a pixel is typically not perceived

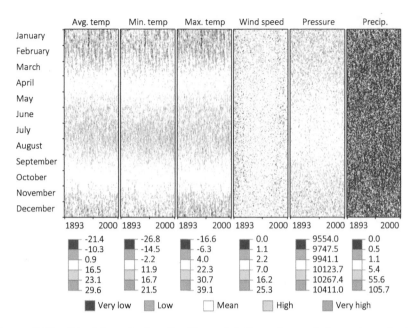

Figure 3.25 Pixel-based visualization of daily values of six meteorological attributes collected for more than hundred years in the city of Potsdam. Courtesy of Thomas Nocke.

in isolation, but together with its surrounding context. As a consequence, our perception is influenced by contrast effects. Therefore, pixel-based approaches should employ color maps with well-discriminable colors.

The second question is about deciding where the pixels of a particular attribute should be shown. A common approach is to partition the display into regions, each being associated with one attribute. Figure 3.25 demonstrates this for meteorological data with 6 attributes. For each attribute, there is a separate rectangular region filled with colored pixels.

Finally, the third question regards the arrangement of pixels within a region. In general, there are different options for arranging the pixels:

- *Element-wise*: The pixels are simply arranged in the same order as data tuples appear in the dataset.

- *Pre-determined*: The pixel arrangement is determined by an independent variable of the dataset. For example, temporal dependencies can be revealed by arranging pixels with respect to time.

- *Attribute-based*: The pixels are sorted according to the data values of a particular attribute. Multivariate correlations can be revealed by such attribute-based arrangements.

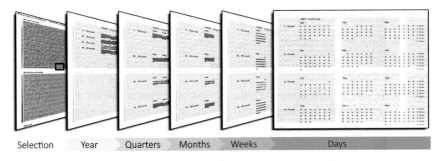

Figure 3.26 Exploding a pixel-based visualization by applying a stepwise separation of a time-oriented recursion pattern. Adapted from [LS07].

- *Query-based*: The pixels are arranged according to the distance of their associated data tuple to a query specification. This way, relevant values (with respect to the query) are grouped and appear at the front of the arrangement.

Our example in Figure 3.25 uses a pre-determined arrangement with respect to time. We can easily see that the first three attributes are strongly influenced by the seasons of the year. The fourth and fifth attributes show a dependency on seasons as well, though less pronounced. The last attribute seems to be independent of time.

Apparently, using pixel as information-bearing primitives is very well suited to create overviews of multivariate data. Yet, details are not so easy to access. One way to remedy this is to use exploded views, which are widely applied in the context of technical drawings. The idea is to add details to complex graphics by separating individual parts, showing these parts by enlarged views, and enriching them with annotations. As illustrated in Figure 3.26, the idea of exploded views can also be applied to pixel-based visualization [LS07]. In the example, a user-selected part of a recursive time-oriented arrangement of pixels is exploded. A stepped animation shows a sequence of "explosions" that transfer the compact and dense pixel selection into a calendar-style representation. Details are now much easier to discern.

3.2.6 Nested Visualization

The final class of techniques for multivariate data visualization to be described in this section is nested visualization. The previous example from Figure 3.26 already involved a nested recursive arrangement of pixels, whereas the other visualization techniques discussed so far had a flat arrangement.

The basic idea of *nested visualization* is to divide the attribute space into subsets and to spatially nest the subsets on the display. This involves two steps:

1. Definition of attribute subsets and nesting order

2. Embedding of subsets within subsets

Definition of Attribute Subsets and Nesting Order   In the first step, three aspects have to be considered. First, the data domain of the attributes must be made compatible for nesting. Nested visualizations require nominal, ordinal, or discrete data domains of small cardinality. Otherwise, a nesting is impractical or cannot be carried out at all.

Second, the attribute subsets have to be specified. All subsets must have the same size. Typically, there are not more than three attributes per subset. It is important to note that different subsets lead to different visual representations, which in turn, communicate different aspects of the data.

Third, the nesting order of the subsets must be defined. In other words, it has to be declared which subset should be nested into which other subset. The nesting order determines which attributes are primary in the visual representation and which are subordinate. Usually, the nesting is rather shallow with hardly more than four levels. The reason is that the deeper the nesting, the harder it is to interpret the visual representation.

Embedding of Subsets within Subsets   Once the attribute subsets and their nesting order have been set up, the next question is how to embed them spatially on the display? Usually, the embedding depends on the size of the attribute subsets. Different techniques exist for subsets with one, two, and three attributes.

Mosaic plots are suited for subsets with only one element, for example $(a_1), (a_2), \ldots$. For the first step, the display area is split along the horizontal axis into rectangular segments. The number of segments corresponds to the number of distinct values of a_1's domain and the size of the segments represents the frequency of the values in a_1's value range. For the second step, each of the rectangular segments is split along the vertical axis with respect to a_2. The procedure of splitting along the horizontal and vertical axes continues for all attributes in the nesting order. At the end, the rectangular segments of a mosaic plot represent the value distribution of the data. Figure 3.27 shows an example with the distribution of surviving passengers of the Titanic with respect to class and sex.

If the nesting order consists of pairs of attributes $(a_1, a_2), (a_3, a_4), \ldots$, then the embedding scheme of *dimensional stacking* can be used [LWW90]. The basic idea is to embed grids within grids. The dimensional stacking starts with dividing the display into a grid of $|a_1| \times |a_2|$ uniformly sized cells, where $|a_i|$ denotes the number of distinct values in a_i's domain. The next step is to subdivide each grid cell into a new grid of dimensionality $|a_3| \times |a_4|$. The procedure continues for all pairs of attributes. In order to actually visualize multivariate data, the grid cells are color-coded according to the frequency of data tuples that exhibit the value combination of the corresponding cell.

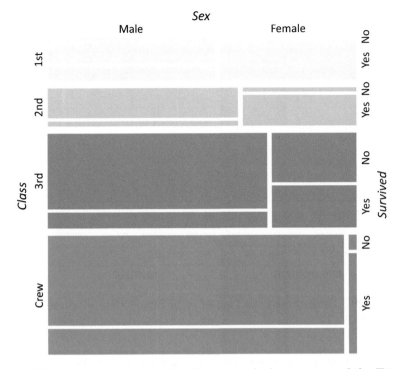

Figure 3.27   Mosaic plot visualizing the survival of passengers of the Titanic.

As an example of dimensional stacking, Figure 3.28 visualizes data about resistance to antibiotics as mentioned earlier. The data consist of eight binary attributes indicating non-resistance or resistance. We can easily see that most of the cells are empty, meaning that no data tuples exist that exhibit the corresponding resistance pattern. We can also easily identify the frequent value combinations indicated in red and the most frequent resistance pattern in the top left corner, which represents no resistance at all.

With mosaic plots and dimensional stacking we have discussed strategies for attribute subsets with one and two attributes, respectively. If the attribute subsets are triples, the *worlds-within-worlds* approach can be used for embedding data into a 3D display space [FB90]. A nesting order with triples could look like this: $(a_1, a_2, a_3), (a_4, a_5, a_6), \ldots$. The three attributes of the first subset define the three axes of a 3D coordinate system. Within this coordinate system, a particular point is fixed by user selection. This point serves as the origin for embedding a new 3D coordinate system spanned by the second triple of attributes. The process of fixing a point and embedding a coordinate system is repeated for all triples.

The coordinate system defined in the last step serves to display the actual data values of the attribute subset at the end of the nesting order. In other words, only three attributes are visualized directly, while the values of the

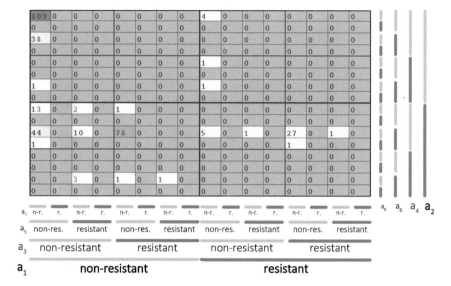

Figure 3.28 Dimensional stacking of eight binary data attributes. Adapted from [SM00].

remaining attributes are expressed by the nesting process. As such, the worlds-within-worlds approach represents only a particular part of the data as specified by the sequence of coordinate origins.

From the previous paragraphs we see that nested visualization of multivariate data is a powerful approach. However, the visual representations are not trivial to construct, and hence, require some training in order to be able to interpret them.

Summary

With the description of nested techniques, we conclude this section on visualization of multivariate data. As we discussed quite different visualization strategies for multivariate data, let us briefly summarize them.

- Table-based visualization transforms common spreadsheets into visual representations. Sorting the table rows allows us to detect multivariate correlations, outliers, and clusters. A focus+context display of the table rows makes it possible to identify individual data values.

- Combined bivariate visualization is suited to communicate two-dimensional value distributions, bivariate correlations, clusters, and outliers. Multivariate relationships can be explored with interactive linking & brushing. Additional visual encodings can be used to represent the frequency of data elements as well.

- Polyline-based visualization shows data elements as polylines across axes. The exploration of two-dimensional value distributions between two neighboring axes is well supported. The analysis of all bivariate relationships in the data requires rearranging the axes. Data frequencies can be visualized by embedding histograms.

- Glyph-based visualization encodes data tuples as small glyphs. The goal is to provide an overview of the data, while details are typically omitted. As such, glyphs facilitate a rough estimation and comparison of the properties of data elements. Glyphs are also suitable for visualization of data in a spatial frame of reference.

- Pixel-based visualization approaches encode each data value by a single color-coded pixel. Thanks to this very space-efficient encoding, even very large datasets can be visualized. Yet, details cannot be discerned easily, unless additional interactive or visual means are employed.

- Nested visualization is based on a recursive embedding of attribute subsets on the display. The resulting visual representations are very well suited to communicate frequencies of the different value combinations in the data. Yet, the interpretation of nested visualizations is not easy, particularly when the nesting is deeper.

In terms of the basic visual design options discussed in the beginning of this chapter, we can conclude the following. The described visualization strategies for multivariate data typically use spatial arrangements in $2D$. There are techniques that use $3D$ spatial arrangements, such as the worlds-within-worlds approach, and also techniques that generate an animated sequence of views, such as the *grand tour*, but in general $2D$ spatial arrangements are much more widely used.

Already at this point, we can further conclude that visualization requires interaction to take full advantage of the power of visual representations. Interaction allows us, for example, to identify data on demand or to explore alternative correlations through rearrangements. In fact, interaction is so important that the entire Chapter 4 is dedicated to this topic.

Moreover, we have mentioned that certain visualization strategies can only be applied when the data domain contains not too many values. We also briefly mentioned the problem of over-plotting when there are very many data elements to be visualized. In these cases, additional computations are necessary to condense the data before they can be transformed into meaningful images. We will discuss this topic in more depth in Chapter 5.

All in all, this section on multivariate data visualization focused on the dependent variables, the attributes A. In the next two sections, we will consider the independent variables time T and space S.

3.3   VISUALIZATION OF TEMPORAL DATA

Time is an exceptional dimension. Virtually everything around us is governed by the steady progress of time. So it comes as no surprise that much of the data that people seek to understand are connected to time. In this section, we add the dimension of time T to our considerations. In other words, we are interested in methods and techniques for visualizing time and temporal data, where the latter primarily means communicating the dependency of data attributes on time $T \to A$. As we will see in a moment, time and data that depend on time are quite special and require dedicated techniques for their visualization.

3.3.1   Time and Temporal Data

Time is not just another data attribute. Time has several properties that need to be taken into account when visualizing data that are connected to time. In the pages to come, we will briefly characterize time and temporal data.

Characterizing Time

Philosophers have pondered the concept of time for ages. Here we want to concentrate on aspects of time that are relevant for the visual analysis of temporal data.

Before going into the details of time, there is a more general aspect to be mentioned: Analog to the different data domains discussed in Section 2.2.1, the time domain can be characterized as ordinal, discrete, or continuous. For *ordinal* time, only equality and partial order relations are defined. For *discrete* time, a mapping exits from the time domain to the set of integers, which makes it possible to measure distances in time. *Continuous* time conceptually maps to the set of real numbers. As such, continuous time is dense, which, in general, complicates the data handling and visualization.

Let us next look at the specifics of time in more detail. In particular, we will deal with four aspects: primitives, arrangement, granularity, and structure of time.

Primitives   In the first place, we have to think about temporal primitives. Temporal primitives serve as anchors to pinpoint certain events or phenomena in time. Two different types of temporal primitives can be distinguished: instants and intervals.

A time *instant* is a singular point in time. An instant is assumed to have no temporal extent. An example of a time instant would be 19:30 o'clock, and we could use it, for example, to agree on a specific time to meet for dinner with friends.

A time *interval* is a temporal primitive with an extent. An interval can be defined explicitly by two time instants: the interval's start and end. Alter-

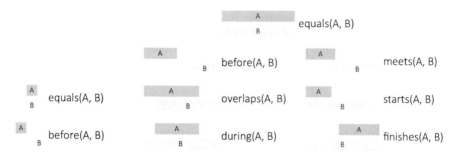

(a) Instant relations.   (b) Interval relations. Adapted from [All83].

Figure 3.29   Temporal relations for time instants and time intervals.

natively, an interval can be defined with a start instant and a duration. For example, when we mark the dinner in our electronic calendar, we may reserve a slot that starts at 19:30 and lasts for 2 hours.

The distinction of temporal primitives is important, because different relations may exist between instants and intervals. As illustrated in Figure 3.29, particularly the relation between intervals can be versatile [All83]. Because understanding temporal relations is a relevant analysis objective, visual representations should be designed so as to enable the observer to identify them.

Arrangement   The arrangement of time basically describes how temporal primitive are laid out in time. The arrangement can be linear or cyclic.

In correspondence with our natural perception of past, present, and future, time can be considered as a *linear* arrangement of temporal primitives. For linear time, we can clearly denominate a temporal primitive as being before or after another one. We may also quantify how much time has elapsed.

Cyclic time is based on recurring temporal primitives. Many phenomena follow the natural cycles on Earth, such as the seasons of the year or the hours of the day. For cyclic time, any primitive is preceded and succeeded at the same time by any other primitive. For example, winter comes before summer, but winter also succeeds summer. A common data analysis objective for unknown data is to figure out whether there are recurring cycles in the data or not.

Granularity   Time is usually measured with respect to a smallest possible unit. If there is only the smallest possible unit, we can say that the time domain has a *single* granularity only. The granularity of time is also closely related to the size of temporal data. The finer the granularity is, the larger we can expect the data to be. A typical example would be simulation data that are given at sub-second granularity.

In many applications, it is practical to consider time with respect to *multiple* granularities, where smaller units are compiled into larger ones. Human-made calendar systems are an example for such multi-granularity time models. They can serve as a natural scaffold for multi-scale data analysis. Studying temporal data at a finer granularity may reveal subtle details, whereas a coarser granularity would convey overall trends or rough estimates.

Structure Finally, the structure of time is relevant [Fra98]. Our regular understanding of time is that it progresses in an *ordered* fashion. There is a single thread of time on which things happen one after the other, and what has happened cannot be changed or undone.

If the exact path through time is unknown, for example, in planning and prediction scenarios, a *branching* time structure unfolds. Each branch describes a possible development in time, of which only one will eventually become true.

If multiple paths through time co-exist in parallel, the conceptual structure of time is denoted as *multiple perspectives*. This type of structure helps, for example, when analyzing eyewitness reports, where each person has their own perspective on the observed reality.

It is obvious that branching time and multiple perspectives require more resources for their visualization than ordered time, simply because more information needs to be encoded visually. The uncertainty inherent in the different time structures has to be communicated as well.

As we see, time is more than a simple linear succession of consecutive moments. It makes a difference if we are dealing with instants or intervals in time, if time is linear or cyclic, if a single or multiple granularities are given, or if time runs in an ordered, branching, or parallel fashion. All of these aspects need to be considered when designing visual representations of time and temporal data. A schematic overview of the discussed aspects is given in Figure 3.30.

After this brief characterization of time itself, we will next take a closer look at how data can be linked to time.

TIME

Primitives	Instant		Interval	
Arrangement	Linear		Cyclic	
Granularity	Single		Multiple	
Structure	Ordered	Branching	Multiple perspectives	

Figure 3.30 Aspects of time to be considered when visualizing temporal data.

Characterizing Temporal Data

In general, temporal data are data with references to time. Depending on whether the values in the data vary with respect to time T, the data are denoted as *temporal* or *non-temporal*. If a dataset as a whole changes over time, that is, there are several versions V of the data, then the data are denoted as *dynamic*. If the dataset is fixed, meaning there is only one version, it is said to be *static*. These denominations lead to a categorization of different types of data:

Static, non-temporal data do not vary over time at all. The well-known Iris flower dataset is an example of data that are completely independent from time.

Static, temporal data can be considered a historical view of how an observed phenomenon evolved during a certain period of time. Common time series are a typical example. They contain time-varying data values, but the dataset itself does not change after it has been created.

Dynamic, non-temporal data are data that change over time, but only a single snapshot of time is available. In other words, we have a continuous stream of data, but without a history. Such data can often be found in monitoring scenarios, for example, when the current state of some machinery is continuously visualized in a control room.

Dynamic, temporal data are data whose values vary over time, and the state of the dataset changes over time as well. Such bi-temporally dependent data constantly evolve and also maintain a history. An example is meteorological data, which contain time-dependent measurements and are updated regularly as new measurements become available.

An illustration of the different types of temporal dependencies of data is given in Figure 3.31. Non-temporal data can be visualized with general techniques for multivariate data as introduced in the previous section. A peculiarity of dynamic data is that the visualization has to keep up with the changes in the data. If dynamic data additionally preserve a history (dynamic, temporal data), often only a small time window of it can be visualized due to the flood of information that needs to be processed.

It should be mentioned that the distinction of the different types of temporal dependencies of data is inspired by database terminology [Ste98]. In the context of visualization, such a clear notation has not yet gained widespread acceptance. Hence, one may find a mix of alternative terms, such as time-varying data, time-series data, or dynamic data, but most of the time the data are in fact static, temporal data.

In the following, we will concentrate on the visualization of such static, temporal data, because this is the class of data that is most relevant in many application domains.

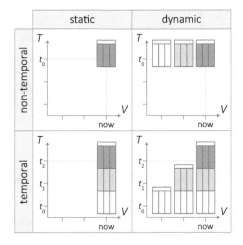

Figure 3.31 Types of data with references to time. Adapted from [Ste98].

3.3.2 Visualization Techniques

In general, any visual variable can be used to encode time and its associated temporal data, where the visual design choices apply as introduced in Section 3.1. A key characteristic of visualization techniques for temporal data is if the visual representation is *static* or *dynamic*.

A static representation does not change over time. It remains fixed and provides an overview of the time axis in a single image. The most common visual encoding is based on *mapping time to space*, which means that the temporal primitives are placed at different locations on the display. A prominent example of a static time-to-space visualization is *small multiples* [Tuf83]. The idea is to visualize the data as a series of miniaturized, high-density, non-overlapping views that are arranged according to time. The views are of equal size and use the same visual encoding, yet each view shows the data for a different temporal primitive. Figure 3.32 illustrates the small-multiples approach. The figure makes clear that static representations can provide a nice overview of the data.

In contrast to static representations, a dynamic representation changes over time. This corresponds to a *mapping of time to time*, more precisely, of time in the data to time of the presentation (or wall-clock time). In other words, the temporal data are shown as an animated sequence of individual images. Dynamic representations can communicate the time-varying behavior of the data very well. However, at any time during a dynamic representation only a single moment of the temporal data is visible, and it is immediately overwritten by the next moment. This can make it more challenging to spot details and to get an overview of the data, particularly when the data are

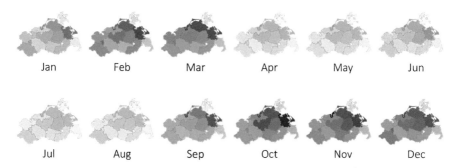

Figure 3.32 Small multiples visualization of the number of people diagnosed with problems of the upper respiratory tract.

larger. Therefore, dynamic representations are usually equipped with controls to pause, rewind, speed up, or slow down the representation.

Knowing about the fundamental distinction between static and dynamic representations, we next introduce concrete visualization techniques for time and temporal data. Because time and temporal data are important in so many applications domains, there exist a plethora of techniques to visualize them. A comprehensive survey is available in specialized literature [Aig+11]. Here, we can only describe a few selected examples, and because this book is a print medium, we concentrate on static representations. To cover a broad range of techniques nonetheless, the examples are organized according to the key aspects of time as discussed before. First, we deal with techniques for time instants and time intervals and then for linear and cyclic time. We will illustrate the utility of multiple granularities and finally take the structure of time into account.

Representing Instants and Intervals

A key difference when visualizing temporal data is whether we are concerned with instants or intervals in time. Instants correspond to points. They bear only a single piece of temporal information. Intervals, on the other hand, correspond to ranges. They bear two pieces of information (start and end, or start and duration). Let us next look at two visualization techniques, one that is suited for instants and one for intervals.

The Time Wheel   In the previous section on the visualization of multivariate data, we introduced axes-based techniques with different layouts of axes and different types of data representations between pairs of axes. The Time Wheel is an axes-based technique particularly for multivariate temporal data [TAS04]. It uses a central time axis around which several attribute axes are arranged in a radial fashion as shown in Figure 3.33.

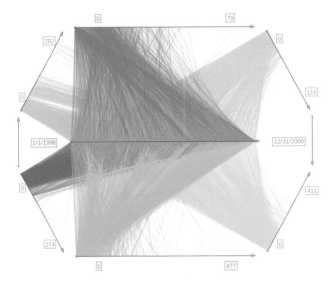

Figure 3.33 Time Wheel visualization of human health data.

The actual data representation is done via lines that connect an instant of the time axis to its corresponding data values at the attribute axes. To this end, time and attribute values, which may be continuous, are projected to their associated axes and a line is drawn between them. As such, the Time Wheel implements a visual encoding based on the screen positions of the lines' start and end points at the axes.

The Time Wheel in Figure 3.33 visualizes temporal data that contain the number of people diagnosed with certain health problems. The time axis uses dates as time instants. Each of the attribute axes shows a different diagnosis. As can be seen, the top and bottom axes can be best interpreted with respect to time. By rotating the attribute axes around the central time axis, it is possible to bring any pair of attributes into this focal position. In its depicted state, the Time Wheel shows moderately low values throughout time, but there are also dates with unusually high numbers of diagnosed people.

Triangular Model   The triangular model is a technique particularly for visualizing intervals [Kul06]. It is based on two coordinate axes, the horizontal one representing time and the vertical one representing duration. In the triangular model, an interval is represented as a dot with two attached arms. The dot is placed so that the arms connect the time axis exactly at the start and the end of the represented interval. The point's height corresponds to the interval's duration. Figure 3.34 provides a schematic illustration.

The triangular model is useful when it comes to reasoning about properties and the relationships of multiple intervals, because it generates easily distinguishable visual patterns for all possible interval relations. There is even

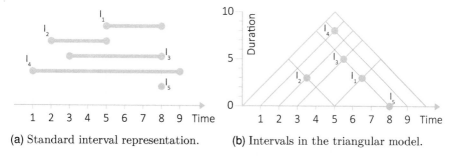

(a) Standard interval representation.   (b) Intervals in the triangular model.

Figure 3.34 Visual representation of intervals using the triangular model. Adapted from [Qia+12].

room for visualizing data that might be associated with the intervals. The dot-based encoding would allow for resizing or coloring the dots based on some attribute values. Yet, the triangular model is only of limited use for multivariate attributes.

Linear and Cyclic Visualization

Different analytic questions can be answered with linear and cyclic visualizations of time. Temporal trends can be communicated well with linear representations. Recurring patterns, on the other hand, are easier to spot with cyclic representations. Below are three example techniques, the first has a linear time axis, the second a cyclic time axis, and the third combines linear and cyclic encoding.

Stream Graph   A stream graph is a technique for visualizing multivariate temporal data with a linear arrangement of time. As in the previous two examples, time is shown along the horizontal display axis from left to right. The multivariate data attributes are visualized as stacked streams, there is one stream for each attribute. The actual visual encoding is based on varying the thickness of the streams along the horizontal axis. That is, the vertical height of a stream at a particular horizontal position represents the underlying data value at the corresponding time. Various alternatives exist for ordering the streams and shaping the overall stack of streams.

As illustrated in Figure 3.35, a stream graph provides a nice overview of the data evolution. On top of that, a stream graph undeniably has an aesthetic value, which is the reason why it has gained widespread popularity [BW08b]. A downside is that individual streams can be difficult to interpret, because we intuitively perceive a stream's thickness as being perpendicular to its flow, whereas the actual visual encoding is always done with respect to the vertical axis irrespective of the ups and downs of the streams.

Figure 3.35 Stream graph with randomly generated data. Adapted from bl.ocks.org/mbostock/4060954.

Spiral Representation   In case cyclic temporal behavior is central to the analysis task, a spiral representation of the data can be created [TS08]. The basic idea is to map the time axis to the shape of a spiral along which the time-dependent data values are shown. The cycle length determines how many values are shown per spiral loop. The data values can be visualized in different ways, for example, by color-coding or by bars of varying length.

We already saw a color-coded spiral at the beginning of this chapter in Figure 3.1. Another example is depicted in Figure 3.36. It shows four years of daily temperature data encoded with the two-tone pseudo-coloring technique. The spiral nicely reflects the recurring seasonal pattern of lower temperatures (blues and greens) in winter and higher temperatures in summer (reds and oranges).

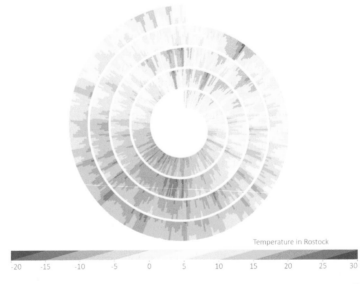

Figure 3.36 Spiral display with four years of daily temperatures in Rostock.

Such cyclic characteristics become visible, only if the cycle length of the spiral matches the length of the recurring patterns in the data. Particularly for data with unknown periodicities it is therefore necessary to enable the user to interactively search for suitable cycle lengths. How the user can be guided in the exploration process will be illustrated later in Section 6.2 of Chapter 6 on advanced visual analysis methods.

In addition to cyclic patterns, the spiral can also communicate linear aspects of the data. By looking along a line from the center to the outer loop of the spiral, we can see how the data evolves from one cycle to the next. In our example from Figure 3.36, we can easily compare summers and winters from the past (inner loops) with the current year (outer loop).

Cycle Plot   The cycle plot is a technique particularly designed for the combined visualization of linear and cyclic components of temporal data [Cle93]. The basic idea is to show the cyclic component as a line plot into which several smaller plots are embedded to visualize the linear component. As such, the cycle plot is a kind of nested visualization.

For the purpose of illustration, Figure 3.37 compares a standard line plot against a cycle plot. The line plot is shown as a reference at the top. The actual cycle plot is depicted at the bottom. Its horizontal axis encodes the

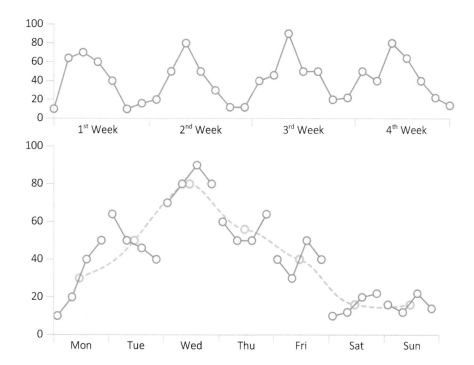

Figure 3.37   Comparison of a regular line plot (top) and a cycle plot (bottom).

days of the week. The dashed line visualizes the average values per day to form the overall representation of the weekly cycle. Per day of the week, we see a smaller plot embedded into the main chart. Each of these smaller plots represents the linear development over four weeks.

The cycle plot allows us to see things at a glance that are not so easy to derive from a standard line plot. For example, it is immediately clear that data values are generally lower on Saturdays and Sundays, and that data values have increased on Mondays, whereas Tuesdays are in a decline.

Considering Granularity

Several of our previous examples visualized temporal data at the granularity of days. Yet, temporal data can be given at different granularities: years, quarters, months, weeks, days, hours, minutes, seconds, or even smaller time units. Next, we look at a visualization technique that considers multiple granularities to create a multi-scale representation of temporal data.

Cluster and Calendar-based Visualization   In order to understand how certain resources (human, energy, computing time, etc.) are utilized over time, it makes sense to collect consumption data and visualize them. Often such data are measured at fine granularity, say minutes or hours. Hence, consumption data can quickly grow to a size that poses a considerable challenge.

Cluster and calendar based visualization addresses this problem by implementing a multi-granularity visualization [vWvS99]. The approach starts out with representing the daily evolution of the consumption. To this end, a classic line plot is used. As shown in the right part of Figure 3.38, the line plot has a granularity finer than hours along the horizontal axis. The vertical axis represents the resource consumption, in our case the number of employees. This kind of representation reveals patterns with respect to the time of the day. But are the patterns the same for all days, or are there characteristic patterns for different days of the year?

To answer this question, the data are clustered. Details on clustering as a general data abstraction strategy will be discussed in Section 5.4.2 of Chapter 5. For the time being, we only need to know that days are grouped with respect to the similarity of the daily consumption curve. A calendar representation as in the left part of Figure 3.38 is then used to visualize cluster affiliation. In other words, the days of the calendar are color-coded according to the cluster to which a day belongs. With this combined representation of calendar and line plot, the user can see the typical pattern of workdays (orange), days that deviate from this regular pattern, for example a series of Fridays during summer vacation (green), and when in the year unusual behavior can be observed, for example, on December 5 or December 31.

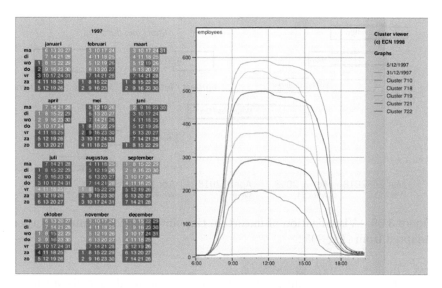

Figure 3.38 Cluster- and calendar-based visualization of temporal data. Reprinted from [vWvS99].

Visualizing Time with a Branching Structure

Last but not least, we said that the structure of time needs to be taken into account. All the examples presented before assume an ordered structure of time. That is, the data are understood as a single unique series of values. Branching time and multiple perspectives are different in that there can be variability in when things happen(ed) in time. Next, we look at a technique that supports the visualization of such variable time structures and also the temporal uncertainty they involve.

Planning Lines   Particularly in the context of scheduling, it is not always certain when activities start or end. The Planning Lines technique has been designed to support planning and clearly communicate the involved uncertainties [Aig+05]. For this purpose, horizontal glyphs encode six pieces of temporal information: the earliest and latest start of an activity, the earliest and latest end of it, and its maximum and minimum duration. A detailed view of a glyph is provided in the upper right box in Figure 3.39.

For the actual planning, multiple glyphs can be placed on a calendar tablature where several activities are allowed to form a branching structure or run in parallel. Dependencies between activities are indicated via links. More precisely, if the start of an activity depends on the successful completion of another activity, the two activities are connected by a link. The links can split and also join paths through time. How a hypothetic planning scenario could look like is illustrated in Figure 3.39.

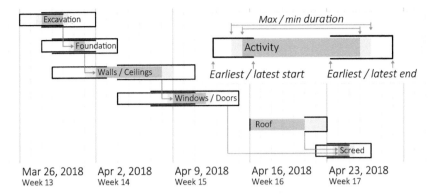

Figure 3.39 Visualization of uncertain time intervals for planning purposes. Adapted from [Aig+05].

Note that the Planning Lines technique is primarily for visualizing the structure of time and uncertainties, and not so much for the temporal data associated with branching time. In fact, visualizing branching time structures together with multivariate temporal data is a challenging problem that requires further research to solve it.

The Planning Lines are the last example in our series of visualization techniques for time and temporal data. The previous pages could cover only a fraction of the rich body of existing work. A comprehensive overview with many more examples is provided by the TimeViz Browser.

The TimeViz Browser

There is a great number of valuable approaches for visualizing time and associated data. A problem is how to find a technique that fits a user's particular needs. The TimeViz Browser aims to provide a solution to this problem.

The idea behind the TimeViz Browser is to provide in a single place a greater overview of what visualizations are possible. The TimeViz Browser is an illustrated survey with a searching and filtering function that allows users to narrow down the scope of techniques according to their needs. To reach a wide audience, the TimeViz Browser is available as a web site accessible at http://browser.timeviz.net. It enables practitioners and researchers to explore, investigate, and compare more than a hundred techniques.

The design of the TimeViz Browser is shown in Figure 3.40. The main view consists of thumbnails and provides a compact visual summary of the available visualization techniques. Selecting a technique opens up the detail view that offers a brief abstract for the technique, a larger figure, and a list of relevant publications. The filter interface to the left covers several criteria, for example,

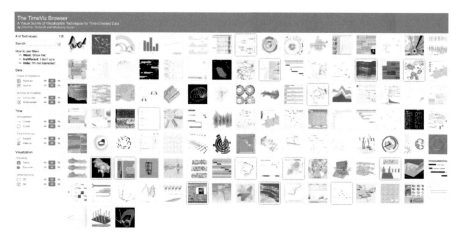

Figure 3.40 The TimeViz Browser provides an illustrated overview of more than a hundred techniques for visualizing time and temporal data.

if a technique is suited for instants or intervals, or for linear or cyclic time. Using the filters it is possible to narrow down the collection of thumbnails presented in the main view, for example, to focus on techniques for a cyclic time arrangement of temporal intervals.

The TimeViz Browser also offers a filter that is related to an aspect that we have left out of consideration so far: the presence or absence of spatial dependencies in the data. In other words, in addition to temporal dependencies of data, there can also be a dependency of data on space $S \to A$, often on the geographic space. This brings us to the next section, where we will discuss the visualization of spatial and also spatio-temporal data in more detail.

3.4   VISUALIZATION OF GEO-SPATIAL DATA

Similar to the role that time plays for understanding data, the space into which the data are embedded is crucial for gaining insight. Given the ubiquity of modern sensory technology, it can be taken for granted that a considerable amount of the data collected today are of spatial nature, typically of geo-spatial nature.

In this section, we are going to visualize data with a geo-spatial frame of reference. Our primary interest will be the communication of the dependency of the data attributes on the geographic space $S \to A$. Tightly connected to this dependency are two of the analytic questions discussed in Section 2.2.2: What is the value at a given position (identification task), and where are the positions with a given value (location task)?

Yet, before we introduce visualization methods that help us answer these questions, we will first look at the aspects being specific to geographic space and geo-spatial data.

3.4.1 Geographic Space and Geo-spatial Data

Geo-spatial data are data that are tagged with a position in geographic space. The probably most well-known geo-spatial data are weather data including measurements of temperature, precipitation, wind speed, and also wind direction. Moreover, traffic data and demographic data are associated with the geographic space around us. The aspects that need to be considered when visualizing geographic space and geo-spatial data will be discussed in the following.

Characterizing Geographic Space

Geographic space is special in that it addresses the spherical shape of planet Earth. Here, we consider two important aspects that characterize the geographic space: dimensionality and scale.

Dimensionality   The geographic space is three-dimensional. A position in geographic space is typically defined by three coordinates: *latitude*, *longitude*, and *elevation*. The latitude is the angle between the equator and the poles. The latitude is 0° at the equator, +90° at the north pole, and −90° at the south pole. The longitude is the angle with respect to the prime meridian in Greenwich. Eastward angles have a positive sign, whereas westward angles are negative. The maximum absolute value of the longitude is 180°. Because the surface of the Earth is not perfectly planar, elevation is used to measure the distance of a geographic position above or below the sea level.

Depending on the application, the two coordinates latitude and longitude may be fully sufficient for the visual analysis of geo-spatial data. For example, visual representations for weather forecasts typically are based exclusively on plotting the data with respect to latitude and longitude. Yet, there are also applications where the elevation part is crucial. Air traffic planning and analysis immediately comes to mind as an example.

Scale   In terms of scale, the geographic space is continuous and infinite with respect to resolution. Essential for data collection, processing, and visualization is the definition of *spatial units* at a certain scale. In the simplest case, the spatial units are points given at a certain spatial scale. Yet, the spatial units may also be given as complex regions at multiple scales, such as federal states, districts, and zip-code regions. In such cases, the spatial units are typically organized in a hierarchical structure (similar to what we said about temporal granularity in the previous section). Figure 3.41 shows the German federal state Mecklenburg-Vorpommern composed of different spatial units at different scales.

The spatial scale significantly affects the results of the visual data analysis. Relationships detected at one scale may not be observable at another scale [And+10]. Hence, determining an adequate spatial scale that corresponds

Figure 3.41 Spatial regions at different scales: Federal state, districts, zip-code regions.

to the subject under consideration is crucial. For example, investigating the traffic at a specific crossing requires a different spatial resolution than exploring major routes between districts.

Often, finding the spatial scale that best matches the task at hand is a trial-and-error procedure. It may even be necessary to create further spatial scales by subsuming or subdividing spatial units. Coarser scales can be derived from the original scale by means of a suitable aggregation strategy. This includes the application of aggregation functions such as average, sum, or count. For the creation of finer scales, a suitable distribution strategy is required to assign data values to the newly specified sub-regions. Usually, additional context information is necessary to arrive at semantically meaningful aggregations and distributions.

Both dimensionality and spatial scale are important characteristics of the geographic space. They have a decisive influence on the results that can be achieved with visual data analysis. Let us next take a closer look at what characterizes geo-spatial data.

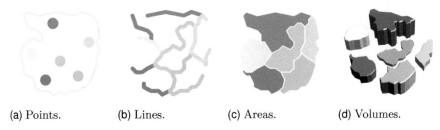

(a) Points.   (b) Lines.   (c) Areas.   (d) Volumes.

Figure 3.42   Geo-spatial data can refer to different spatial units.

Characterizing Geo-spatial Data

Geo-spatial data (also called spatial data or geographic data) have explicit references to the geographic space. In other words, the data are associated with spatial units. As illustrated in Figure 3.42, there are four types of spatial units: points, lines, areas and volumes. For example, data collected at a measuring station are associated with the very position of the station. Highway traffic data would refer to lines. Census data typically refer to discrete administrative spatial areas. Weather phenomena such as thunderstorms are an example where geo-spatial data have a volumetric anchor in space.

It is the existence of a relationship between the data and spatial units in geographic space that characterizes geo-spatial data in the first place. The relationship itself is further characterized by the *first law of geography* as formulated by geographer and cartographer Waldo R. Tobler [Tob70]:

"[...] everything is related to everything else, but near things are more related than distant things."

Tobler, 1970

This law has immediate consequences for the visualization of geo-spatial data. The law's first part implies that geo-spatial data can be interpolated and extrapolated. The second part tells us about the weighting of near and distant data points. Pictorially speaking, the visual representation of a geo-spatial data value can spread in a certain neighborhood. The scope of the data, as explained in Section 2.2.1 of Chapter 2, defines how far the data may spread.

At this point, we know about the specific character of the geographic space and geo-spatial data. In addition to these specifics, the characteristics of multivariate data apply as described in Section 2.2.1. Next, we will introduce strategies for visualizing the geographic space and the associated data.

3.4.2 General Visualization Strategies

Visualizing geo-spatial data means showing univariate or multivariate data attributes A in relation to the geographic space S. This allows us to investigate spatial dependencies $S \times A$. If the data have an additional temporal dimension T, a spatio-temporal visualization can be created to support the analysis of spatial and temporal dependencies $S \times T \times A$. A prerequisite for geo-spatial data visualization is the representation of the geographic space.

Representation of Geographic Space

Geographic space is typically represented by 2D maps. Maps are universally useful. They allow us to develop an understanding of the world and its natural or manmade phenomena. Cartography is the discipline dealing with the production of maps. The cartographic literature provides a wide range of design concepts, guidelines and conventions for creating maps [Mac95]. One such convention is to use blue colors for water, green for lowlands, and brown and white for mountains and their icy caps.

An essential step when creating a map is to project the spherical geographic coordinates to the planar canvas. There are various projection methods, some of which even date back to ancient times. The difficulty that all projection methods have to deal with is to find a trade-off between preserving several properties of the Earth's surface, including area, shape, direction, and distance. For example, the Equirectangular projection, attributed to Marinus of Tyre, who lived around 70–130 CE, preserves distances. The famous Mercator projection, which preserves angles and shapes, was invented in 1569. Even today, new projection methods are devised, for example the Natural Earth projection by Tom Patterson from 2011. Figure 3.43 illustrates these three projections. A particularly interesting class of projections is myriahedral projections, which

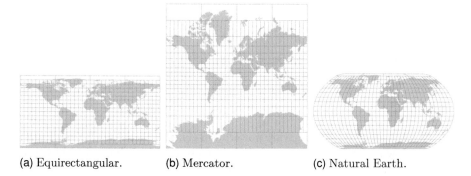

(a) Equirectangular.   (b) Mercator.   (c) Natural Earth.

Figure 3.43 Different map projections preserve different spatial properties. Adapted from `bl.ocks.org/mbostock/3757119`, `.../3757132`, and `.../4479477`.

Figure 3.44 Myriahedral projections of the Earth. Adapted with permission by Nicolas Belmonte from `philogb.github.io/page/myriahedral/`.

unfold the globe by cutting and folding the faces of polyhedra [vWij08; BW18]. The three examples in Figure 3.44 illustrate the new perspective on planet Earth offered by these projections.

Another important step when creating maps is cartographic generalization. It addresses the issues of spatial scale as indicated before. Cartographic generalization aims at reducing the complexity of the visual representation of spatial features by depicting them at lower resolution. The difficulty is to preserve both the geometric characteristics (graphic generalization) and the essential data characteristics (semantic generalization). Figure 3.45 gives an example where the map of Mecklenburg-Vorpommern, as shown earlier, is displayed at different resolutions. The map data have been reduced with the classic line generalization algorithm by Visvalingam and Whyatt [VW93]. At 50% reduced data size, the visual representation of the map is almost identical to the original data. At 10%, the reduced map resolution is fairly obvious, but still the major features of the map can be discerned.

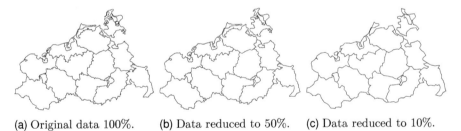

(a) Original data 100%.   (b) Data reduced to 50%.   (c) Data reduced to 10%.

Figure 3.45 Map representation at different resolutions. Generated with `mapshaper.org`.

While 2-dimensional maps are powerful in general, there are also application scenarios where it is necessary to faithfully represent the 3-dimensional features of the geographic space. Sophisticated terrain rendering techniques have been developed to create almost photo-realistic 3D images of the Earth's surface. Terrain rendering is a complicated matter in itself and involves several steps including the generation of 3D meshes, definition and selection of appropriate resolutions, hardware-accelerated shading and texturing, and advanced post-

4392m

0m

Figure 3.46   Terrain rendering of the Puget Sound region. Courtesy of Steve Dübel.

processing. Here, we will not go into the details of terrain rendering, but instead refer the interested reader to the dedicated literature on the topic [Ruz+12].

An example of a 3D terrain rendering of the Puget Sound region is shown in Figure 3.46. Color is used to visualize the terrain elevation. In the following pages, we will see further examples, where geo-spatial data are visualized in 3D scenes. Yet, before that, we will introduce basic visualization strategies for geo-spatial data.

Visualization of Geo-spatial Data

The essential task when visualizing geo-spatial data is to convey the relationship between the geographic space S and the data attributes A. In particular, we will concentrate on two key questions: First, how to combine S and A visually, and second, whether to draw S and A in 2D or 3D?

Direct and Indirect Visualization   There are two baseline options for the visual combination of S and A. We can directly show the geo-spatial data within the presentation of the geographic space. This shall be denoted as *direct visualization*. Alternatively, we can pursue an *indirect visualization*. In this case, the geo-spatial data and the geographic space are depicted in separate views, and their connection is communicated indirectly via visual cues.

Direct Visualization   For direct visualization, geo-spatial data are embedded directly into the visual representation of the geographic space. A prominent example is choropleth maps, for which the spatial units of the geographic space are color-coded according to their associated data. While such choropleth maps are intuitive and easily understood, they usually depict only one or two geo-spatial attributes.

Multivariate geo-spatial dependencies can be communicated by placing glyphs, as described in Section 3.2.4, directly on the map. To this end, suitable glyph positions must be determined. Ideally, glyphs should be located directly at their associated spatial unit. At the same time, glyphs

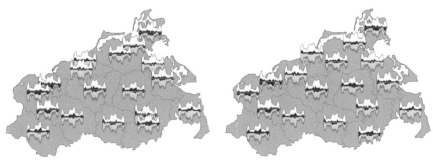

(a) Straightforward placement.   (b) Overlap-optimized placement.

Figure 3.47   Reducing overlap of stream graph glyphs on a map.

should neither overlap each other nor occlude important geographic features. These requirements make glyph placement a non-trivial problem whose solution may involve sophisticated optimization algorithms [FS04]. Figure 3.47 illustrates the difference between a straightforward and an overlap-optimized placement of stream graph glyphs.

Overall, direct visualization allows us to easily identify the data values at certain positions and to locate positions with certain data values. However, with an increasing number of spatial units and data values, direct visualization reaches its limits, because it becomes increasingly difficult to embed all geo-spatial data directly into the map. This is where indirect visualization can help.

Indirect Visualization For indirect visualization, geo-spatial data and geographic space are represented in distinct views that are linked through visual cues. For example, Figure 3.48a shows a choropleth map and a parallel coordinates plot. The map visualizes a single attribute by color for the entire geographic space. The parallel coordinates plot shows all five attributes for all data tuples. Yet, the connection between geographic space and the multivariate data is possible only for a single location, which is marked in the map via cross-hairs, and whose associated data tuple is highlighted in red in the parallel coordinates view.

One way to increase the number of locations for which multivariate geo-spatial data can be visualized is to use geo-spatial probes [But+08]. The goal of probes is to achieve a more flexible combination of the map and several views showing the geo-spatial data. To this end, the user can place a number of probes on the map. For each probe, a separate view is created that shows the data associated with the space around the probe. The connection between the probed spatial locations and the data views is established via visual links. A probe-based indirect visualization is illustrated in Figure 3.48b.

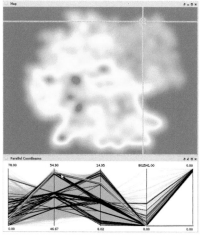 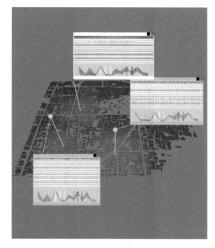

(a) Univariate choropleth map plus multivariate parallel coordinates plot.

(b) Flexible visualization via probes. Courtesy of Thomas Butkiewicz.

Figure 3.48   Indirect visualization of geo-spatial data.

A major advantage of indirect visualizations is that we can show larger amounts of geo-spatial data. Yet, the user must mentally connect two or more views to explore the data's spatial dependencies.

In fact, both direct and indirect visualization are viable options for communicating geo-spatial data. When the spatial aspect S is in the focus, direct visualization tends to be better suited. When the data attributes A play a stronger role, indirect visualization can be more practical.

Let us next continue with the question of whether geographic space and geo-spatial data should be depicted in 2D or 3D.

2D and 3D Visualization

We already discussed the general design decision of using 2D or 3D presentations in Section 3.1.2. When visualizing geo-spatial data, this decision is more intricate, because we have to distinguish between the presentation of the geo-spatial data A and the presentation of the geographic space S [Düb+14].

A systematic view on this concern is given in Figure 3.49. The vertical axis schematically depicts the difference between 2D and 3D presentations of the geo-spatial data, while the horizontal axis compares 2D and 3D presentations of the geographic space (map or terrain). The four possible combinations of 2D and 3D can be characterized as follows.

2D Data Visualization on 2D Map Showing both data and space in 2D is the quasi-standard for geo-visualization. Many established techniques

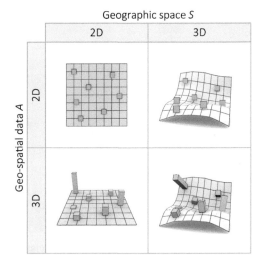

Figure 3.49 Systematic view of 2D and 3D representations of geo-spatial data and geographic space. Adapted from [Düb+14].

belong to this category, including choropleth maps, contour maps, dot maps, flow maps, cartograms, or glyph maps, as shown earlier in Figure 3.47. The level of visual abstraction achieved by purely 2D approaches usually makes the resulting images easy to interpret.

2D Data Visualization in 3D Terrain In this category, 2D and 3D components are combined. A 3D terrain rendering represents the geographic space in its full extent. The geo-spatial data are mapped onto the 3D terrain via 2D billboards or 2D textures, as already illustrated in Figure 3.46. This makes it possible to study how the data behave in relation to specific landscape characteristics, such as mountains or valleys.

3D Data Visualization on 2D Map For this category, the geographic space is abstracted as a 2D map, whereas 3D graphical elements are used for the data. In a sense, the third dimension is exploited for a richer depiction of the data, for example for showing more data values, providing clearer arrangements, or incorporating more dimensions, for example the dimension of time, as we will see a little later.

3D Data Visualization in 3D Terrain If three-dimensional spatial relationships play an important role for the data analysis, it makes sense to show both data and space in 3D. An example is given in Figure 3.50. The visualization shows how an aircraft decreases its speed (color changes from dark blue to bright blue) while maneuvering through the mountains near Sion airport during the approach.

Figure 3.50 3D visualization of the trajectory of an aircraft approaching Sion airport. Courtesy of Steve Dübel.

In general, both 2D and 3D representations are suited for the visualization of geographic space and geo-spatial data. Which combination to choose depends again on the specific data and tasks.

In the case that 3D components are involved in the visualization, it is definitely necessary to address the problems due to occlusion and perspective distortion. In the first place, it is important to make the user aware of the fact that information may be obscured. This awareness enables the user to act and resolve potential problems by changing the viewpoint on the 3D scene.

Visibility widgets as illustrated in Figure 3.51 can help the user in this regard [RS17]. The panoramic view in the bottom right corner allows the user to look with an extremely wide angle onto the 3D scene. The central circular lens visualizes the data otherwise hidden on the backside of the terrain. The color-coded bands at the bottom and to the left indicate where and at

Figure 3.51 Visibility widgets help users identify obscured information in 3D geo-visualizations. Courtesy of Martin Röhlig.

what distance the majority of obscured information is located. While such sophisticated tools might not always be available to the user, it is important to provide at least rudimentary support for an effective exploration of 3D visual representations.

In summary, direct and indirect visualization, as well as 2D and 3D visualization, are the basic options when designing visual representations of geo-spatial data. Next, we will put these basic strategies to use for the combined visualization of data in space and time.

3.4.3   Visualizing Spatio-temporal Data

So far, we considered the visualization of geo-spatial data and temporal data separately. Yet, space and time are tightly connected. Spatial phenomena often include a temporal aspect, and evolving phenomena are often embedded in space. This leads us to the question: how to visualize *spatio-temporal data?*

Spatio-temporal data have references to both space S and time T, a fact that makes the visual data analysis more difficult. It is no longer sufficient to understand in isolation how data values are distributed in space $S \to A$ and how they evolve over time $T \to A$. It is rather necessary to combine both perspectives to enable a comprehensive understanding of spatio-temporal dependencies $S \times T \to A$. Designing corresponding visual representations is challenging. Particularly for the case of spatial-temporal data visualization, it is important to balance the designer's wish to communicate a lot of information and the users' capacity to conceive it.

Probably the most widely used solution is to show an animated map where the visualization changes with each frame. We also mentioned *small multiples* as a suitable approach and gave an example in Figure 3.32 on page 87. In general, we could combine and link two dedicated views, one visualizing the temporal aspect of the data and the other showing the spatial aspect. However, in all these cases, it would be up to the user to mentally integrate S and T.

In the following, we want to illustrate how 3D data visualization on 2D maps can help us obtain visual representations where S and T are already integrated. Our first example will be about the visualization of spatio-temporal movement data. The second example is concerned with the visualization of spatio-temporal health data.

Stacking Movement Trajectories in 3D

Movement data capture how some observed objects move through space and time. Here, we consider movements where each data point is defined as a tuple with latitude and longitude coordinates, a timestamp, and several data attributes, including speed, acceleration, or sinuosity.

The standard way of looking at such data is to study 2D maps where movement trajectories are visualized as 2D paths. Figure 3.52a shows an example with trajectories of cars moving along roads. Color is used to encode

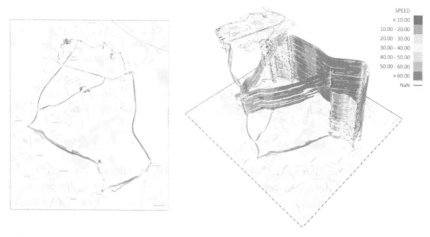

(a) 2D map with 2D paths.   (b) 2D map with stacked 3D bands.

Figure 3.52   Visualization of movement trajectories.

the cars' speed. This basic form of representation suffers from overplotting which makes it difficult or even impossible to see any details of the spatio-temporal behavior of individual trajectories.

At this point, the third display dimension can help us untangle the situation. As illustrated in Figure 3.52b, the simple, yet powerful idea is to tilt the map in 3D and to stack 3D trajectory bands along the third dimension, where each band is assigned a separate z-coordinate [Tom+12].

The 3D trajectory visualization has two advantages. First, we get a comprehensive overview of all trajectories. Second, it is now possible to investigate each individual trajectory in detail. The spatial aspect can be studied in relation to the base map. For the selected trajectory, a dynamic auxiliary map layer is displayed. It eases the interpretation of the spatial embedding, particularly for the data at the top of the stack. Moreover, the bands in the stack can be grouped according to their similarity in terms of their shape. This way, we can see quite well *where* in space the data exhibit certain characteristics, for example, where the speeds are higher (green) and where they are lower (red). But what about the *when* as captured by the temporal aspect of the data?

Again, the third dimension comes to our help. In contrast to 2D paths, the 3D bands offer enough space to include arrows that indicate the direction of the movement. This allows us to see the temporal sequence of places visited during a movement. Moreover, the stacking order can be altered with respect to time. Trajectories from the past could be put at the bottom of the stack, whereas the most recent trajectories could go to the top, or vice versa. Such a temporal ordering of the trajectory bands can reveal long-term changes in the overall movement data.

Both the arrows in the bands and the ordering of the stack allow us to draw qualitative conclusions about temporal dependencies. We can tell that one data point is temporally before or after another one. But quantitative statements about the temporal distance between data points cannot be made. In Section 4.6.3 of Chapter 4, we will learn how an interactive lens technique can help us remedy this problem for movement data. Next, we will continue with an approach where time is granted more visual prominence by design.

3D Space-time Cube Visualization

Under the assumption that the geographic space is defined by two coordinates, spatio-temporal data can be mapped directly to the three display dimensions. This general visualization approach is called a *space-time cube*. The x-axis and the y-axis of the cube are used to map the two spatial coordinates, the third z-axis shows time.

The space-time cube (STC) has been developed long before visualization became an independent field of research [Häg70]. The primary purpose of STC visualizations is to show the paths that objects have taken through space and time [Kra03]. In the following, we will use the STC concept to visualize spatio-temporal health data. The spatial frame of reference will be visualized as a basic 2D boundary map. The actual data attributes, in our case the number of people diagnosed with certain diseases, will be visualized via different 3D graphical representations.

(a) Pencil glyphs for linear trends.   (b) Helix glyphs for cyclic patterns.

Figure 3.53   Visualizing spatio-temporal data using 3D glyphs on a 2D map.

3D Glyphs in a Space-time Cube   Let us start with 3D glyphs. As in 2D, glyphs are created and placed at a suitable location for each spatial unit of the map. Figure 3.53 shows 3D pencil glyphs and 3D helix glyphs [TSS05]. Both types of glyphs are oriented along the z-axis, which represents time. Colored glyph segments visualize individual data values.

Pencil glyphs facilitate the visual analysis of linear temporal trends. Mapping different attributes to the different faces of the pencil allows us to see correlations in the evolution of multiple data attributes. The example in Figure 3.53a shows the monthly number of people with different infections of the upper and lower respiratory tracts.

Helix glyphs emphasize cyclic patterns in the data. A helix glyph is basically a 3D band that winds up along the z-axis. Sub-bands visualize different data attributes. Figure 3.53b shows the same data as Figure 3.53a. Each helix glyph is configured to show 12 months per helix loop with the seasonal peaks currently facing toward the viewer.

Both pencil and helix glyphs have the advantage that spatial and temporal aspects are shown within a single STC image, which allow us to study multivariate spatio-temporal dependencies. Yet, while the temporal evolution of the data is nicely visualized along the individual glyphs, the spatial evolution is more difficult to extract. Because of the empty space between glyphs, the user has to mentally link the information displayed in one glyph to what is shown in another glyph. How an alternative visual representation can resolve this problem will be explained next.

3D Wall in a Space-time Cube   The question is, how can we avoid gaps in the visual representation of the data so that the spatial characteristics are easier to interpret? An interesting solution is to create a non-planar slice through the three-dimensional space-time continuum onto which the spatial-temporal data can be projected [TS12].

The creation of the slice is based on three steps as shown in Figure 3.54:

1. Define topological path through neighborhood graph

2. Construct geometrical path through spatial units

3. Extrude slice for data visualization

In the first step, a topological path has to be defined through the neighborhood graph of the spatial units. The topological path guarantees that the data visualization is free of gaps. In a second step, the topological path is transformed into a geometrical path. This path is constructed in the x-y-plane taking into account the geographic characteristics of the spatial units. A good geometrical path should have low curvature and follow the spatial units' shapes without passing through other territories. Finally, the geometrical path is

(a) Topological path.   (b) Geometrical path.   (c) Extruded slice.

Figure 3.54   Creation of a non-planar 3D slice through space-time.

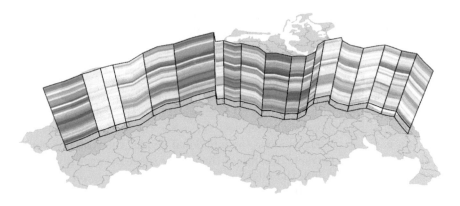

Figure 3.55   Spatial-temporal visualization along a wall on a map.

extruded in the direction of the z-axis to form a non-planar slice with a wall-like 3D shape. This wall acts as a kind of canvas onto which we can project visual representations of the spatio-temporal data.

Figure 3.55 shows an example with the health data that we have already visualized via glyphs earlier. Per wall segment, colors visualize the number of sick people for 36 months of observation. As there are no gaps in the visualization, it is now easier to follow the data's evolution through space and time. However, we are restricted to the path defined in the first place. Therefore, it is important to provide flexible means to adjust the path interactively or based on the characteristics of the data. For example, data-driven paths could be defined by pursuing a gradient descent along spatio-temporal trends in the data.

In conclusion, we see that designing visualizations of spatial and spatio-temporal data involves numerous design choices which all must be weighed and decided carefully. The primary goal must be to clearly represent the spatial dependencies of the data. This goal can be achieved on the basis of suitable 2D map or 3D terrain representations. The actual data representation can be done in a variety of ways, where different combinations of 2D and 3D graphical elements are possible. Conceptually, the data can be visualized directly in the spatial frame of reference or indirectly as a dedicated view that is linked with the map or the terrain.

When the dimension of time is added to the visualization, matters become still more complex. We have seen that using the third display dimension is a viable option to integrate time. Yet, when time, space, and multivariate data attributes are to be analyzed in concert, we usually have to make compromises. In our last example, we used a wall-like 3D visualization without gaps to ease the spatial interpretation of the data, but the solution shows only a single data attribute. 3D glyphs are capable of visualizing multiple spatio-temporal attributes, but some information may be invisible at the glyphs' back face. Because it is necessary to make compromises, a comprehensive visual analysis of spatio-temporal data should ideally provide the user with different visualization techniques and easy ways to configure them.

Next, we will learn that the visualization of graphs, which typically do not have a natural embedding in space or time, is a considerable design challenge as well.

3.5   GRAPH VISUALIZATION

In the previous sections, we have learned how to visualize multivariate data attributes A, the temporal context T, and the spatial frame of reference S. Last but not least, this section will deal with the visualization of relations R among the data. Following the same pattern as in the previous sections, we will first look at the specifics of the underlying data model and then introduce dedicated visualization solutions and illustrate them with selected examples.

3.5.1   Graph Data

The data model that interests us now is graphs. A graph $G = (V, E)$ consists of a set of vertices V (or nodes) and a set of undirected or directed edges E (or links) between the nodes. Roughly speaking, nodes represent pieces of data, whereas edges correspond to the relations among the data. We can also say that the edges define a *structure* among the data.

Networks and Trees

Graphs may come in a variety of different forms. Depending on the characteristics of the structure of a graph as defined by its set of edges, one can distinguish different classes of graphs. If there are no particular constraints imposed on the graph, it is commonly called a *network*. As such, networks capture the intuitive notion of binary relations between entities. Social networks and computer networks are prominent examples. But also transportation systems and biological systems are often modeled as networks.

In contrast to basic networks, a *tree* is a graph that obeys specific constraints. In particular, a tree is a connected, acyclic graph. Connected means that for any two nodes in the graph a path exists connecting the two nodes. Acyclic means that the graph does not contain paths where start node and end node are the same. Moreover, being an acyclic graph also implies that the path that exists between any two nodes is unique.

As in nature, a tree's root and leaves are special. A tree with a designated *root* node $r \in V$ is called a rooted tree. The root serves as the center of the tree from which edges are directed away describing cascades of parent-child relations. A node that has no edges to child-nodes, but only a single edge to its parent-node is called a *leaf*. A managerial hierarchy of a company is an example for a rooted tree with the CEO as the root, and team leaders being the leaves.

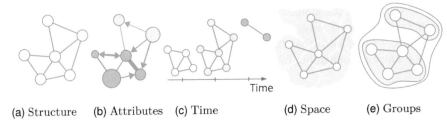

(a) Structure   (b) Attributes   (c) Time   (d) Space   (e) Groups

Figure 3.56   Facets to be considered when visualizing graphs. Adapted from [HSS15].

Networks and trees are the most commonly encountered graphs in the context of visualization. There are other, less common graphs. For example, in bipartite graphs, the nodes can be partitioned into two sets, and edges run only between the two sets, not within the sets. For another example, hypergraphs are graphs where an edge is allowed to connect more than two nodes. Many more types of graphs are known in the literature [BLS99; GYZ14]. Yet, in this book, we want to focus on the visualization of networks and trees.

Facets of Graphs

Figure 3.56 provides an overview of the facets that are most relevant in the context of graph visualization. The *structure* in Figure 3.56a is a primary facet. Yet, additional facets can also play a role when analyzing graph data. These additional facets lead to distinct types of graph data:

Multivariate graphs have additional data *attributes* associated with their nodes and/or edges as illustrated in Figure 3.56b. The data attributes provide additional information about nodes and edges, for example, importance or weight.

Dynamic graphs vary over *time*. As shown in Figure 3.56c, for each point in time, the set of nodes and the set of edges can be different. Between subsequent time steps, nodes and edges may continue to exist, leave the graph or enter it.

Spatial graphs are associated with coordinates in *space* as in Figure 3.56c. The coordinates define the layout of nodes and sometimes even the routing of the edges. A network of flights connecting airports is an example of a spatial graph.

Compound graphs partition the nodes and the edges into *groups* that are typically nested hierarchically, which is depicted in Figure 3.56e. The groups can be expanded or collapsed to create views of the graph with different levels of abstraction.

As mentioned before, the structure of graphs is fundamental, while attributes, time, space, and groups may provide additional context information to complete our understanding of the data. Yet, these additional facets also make the visualization more difficult to design. Therefore, we will start simple and introduce basic visual representations for the structure of graphs in the next section. Later in Section 3.5.3, we will discuss how the additional facets can be incorporated.

3.5.2 Basic Visual Representations

There are three fundamental categories of graph representations: node-link representations, matrix representations, and implicit representations. Additionally, there are hybrid representations, which are basically combinations of the fundamental representational paradigms.

Node-link Representations

Node-link representations (or simply node-link diagrams) depict nodes as dots and the edges between them as lines or arcs. This representation is by far the most common one for networks and trees. A key question is, where should nodes and edges be drawn? Solving this problem is the task of graph layout algorithms [Bat+99; Tam13].

The literature defines a number of conventions, aesthetic criteria, and constraints that graph layouts should obey and fulfill. For example, nodes should not overlap and the number of edge crossings should be minimal. Curved edges may be aesthetically more pleasing than straight lines or orthogonal edges. In general, it is not possible to observe all of these requirements, as they may be contradictory to each other. Therefore, the graph layout problem is typically formulated as an optimization task.

Depending on the degree of freedom that an algorithm has to find suitable node positions, three classes of layouts can be distinguished [SS06]. There are:

- free layouts,

- fixed layouts, and

- stylized layouts.

For a free layout, there are no particular constraints with respect to the node positions. Such layouts are typically generated with force-directed approaches. The goal is to place connected nodes near to each other without cluttering them in a single spot. To this end, force-directed algorithms simulate repulsive forces between all nodes and attractive forces between adjacent nodes [FR91]. With an adequate configuration and a suitable number of iterations, the simulation should converge in an equilibrium that represents the final layout.

In contrast to free layouts, *fixed layouts* are based entirely on predefined node positions. Only the edges can be routed flexibly by the visualization

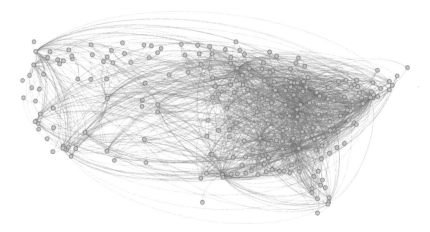

Figure 3.57 Node-link diagram of flights connecting US airports. Created with `gephi.org`.

method. Fixed layouts are typically encountered when the graph is embedded into some spatial frame of reference. Figure 3.57 shows an example where nodes are placed according to the positions of US airports.

In between free and fixed layouts, there are the so-called *stylized layouts*. On the one hand, they are not as restricted as fixed layouts, and on the other hand, they are not entirely free. A typical scenario is that possible node positions are confined to a certain predefined scheme. For example, the nodes may be required to be located on a circle or be arranged in an axis-aligned fashion. Tree layouts often aim to place the nodes of individual hierarchical levels on distinct horizontal or vertical lines.

Once a layout has been computed, it can be post-processed to improve its appearance and readability. For example, the graphical representations of the nodes may overlap, because many layout algorithms neglect the fact that space is needed to actually draw a node. Therefore, node overlap removal is a commonly applied post-processing step [Nac+16]. Another option is to route edges in compact bundles. The bundling of geometrical primitives is a general approach to de-clutter visual representations. It will be explained in more detail in Section 5.1.2 of Chapter 5.

Overall, node-link representations are well suited to display graph nodes and edges in a balanced way. Their widespread use testifies to the universal utility of node-link diagrams. Yet, for graphs with very many edges, node-link representations can get cluttered so much that they resemble hairballs prohibiting any visual analysis. Moreover, the time complexity of optimizing the layout can be high. Matrix representations deal time complexity for space complexity and avoid the edge clutter as we will see next.

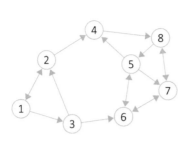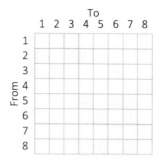

Figure 3.58 Node-link diagram and corresponding matrix representation. Adapted from [vHSD09].

Matrix Representations

Matrix representations are a graphical interpretation of a graph's adjacency matrix. There is exactly one row and one column for each node $v \in V$. If there exists an edge $(v_i, v_j) \in E$ between the i-th and the j-th node, the corresponding matrix cell at position $[i, j]$ (and $[j, i]$ for undirected graphs) is marked. Otherwise, the cell remains empty. The marks placed in the cells can also be utilized to visualize the edge weight or other edge attributes, for example, by varying the color or the size of the marks.

Figure 3.58 depicts a simple node-link diagram and a corresponding matrix representation. As we can see, a matrix is visually focused on the graph's edges, whereas nodes are merely represented as labels. This makes it possible to detect structural patterns such as those illustrated in Figure 3.59.

However, such patterns appear only if the nodes are ordered appropriately along the horizontal and vertical axis. There are many ways to order the nodes of matrices [Beh+16]. A simple approach is to sort them according to a given

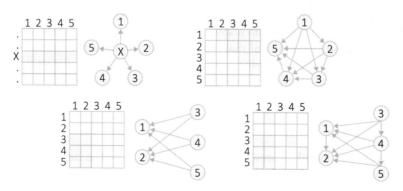

Figure 3.59 Graph patterns represented as matrices and node-link diagrams. Adapted from [SM07].

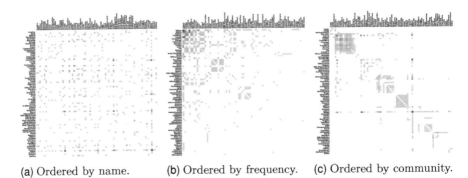

(a) Ordered by name.　　　　(b) Ordered by frequency.　　(c) Ordered by community.

Figure 3.60 Differently ordered matrix representations of the same data. Adapted from `bost.ocks.org/mike/miserables/`.

node attribute. More complex approaches cluster the nodes based on structural properties of the graph so that similar rows and columns are placed next to each other. As finding the *right* ordering that reveals the characteristic features of the graph remains challenging, interactive rearrangement is often helpful.

Figure 3.60 illustrates the impact that the ordering has. All three matrices show the same co-occurrence graph of the play *Les Misérables*, which we already visualized in the introduction of this book in Figure 1.1 on page 7. Different shades of green visualize the number of co-occurrences of two characters, where darker greens represented larger numbers. The shades of blue in the matrix diagonal visualize the overall frequency of a character's occurrences. Three different ordering procedures are depicted. In Figure 3.60a, rows and columns are simply ordered by the name of the characters. The matrix in Figure 3.60b has been ordered according to the characters' occurrence frequency. Finally, Figure 3.60c shows the matrix ordered by the result of a community detection algorithm. As we can see, the more sophisticated the ordering procedure is, the clearer are the patterns in the matrix representation.

Taken together, the main benefit of matrix representations is their focus on the graph's edges, whereas the nodes are merely indicated as labels along the rows and columns. The visual mapping of the graph data is straightforward. While there is no need for layout optimization, matrix ordering is an intricate problem. It should further be noted that a matrix's drawing space grows quadratically with the number of nodes.

Implicit Representations

Matrix and node-link representations as described before have in common that they visualize the relations in the data, that is, the edges by means of dedicated graphical primitives (marked cells or links). In this sense, we can call these representations *explicit*.

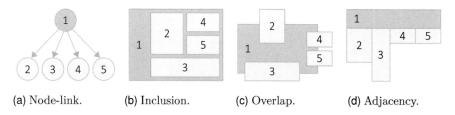

(a) Node-link.　　　(b) Inclusion.　　　(c) Overlap.　　　(d) Adjacency.

Figure 3.61 Node-link representation compared to implicit representations. Adapted from [SHS11].

In contrast to that, *implicit* representations lack explicitly drawn edges. Instead, relations between nodes are encoded implicitly by the relative position of the nodes. However, this strategy is applicable only if the graph structure follows some regularities. General networks can hardly be visualized by implicit approaches. Yet, implicit representations are very well suited for trees with their regular parent-child relations.

There are three options for the implicit encoding of parent-child relations. The encoding can be based on inclusion, overlap, or adjacency as illustrated in Figure 3.61. For inclusion, the child nodes are included in the parent node. For overlap, the children merely overlap their parent. For adjacency, children and parent are placed next to each other.

The absence of explicitly drawn edges obviously puts a visual emphasis on the nodes. In fact, implicit representations often aim to be *space-filling*, meaning that they strive to place the graphical primitives representing the nodes as tightly as possible in the available drawing space. Depending on the applied layout strategy, this can be fast and simple, or time-consuming and complex.

Examples of implicit representations are abundant [SHS11]. Figure 3.62 shows a few selected techniques, all visualizing the same classification hierarchy of mammals with about 3,000 nodes. The 2D squarified treemap in Figure 3.62a is based on inclusion [BHvW00]. The information pyramids in Figure 3.62b use adjacency to encode the parent-child relations by stacking the levels of the hierarchy along the z-axis [AWP97]. The 3D sunburst in Figure 3.62c uses adjacency as well, but now the levels are organized as a kind of shells around the central root node [SHS11]. For many more examples, the reader is referred to the website http://treevis.net, which collects all kinds of tree visualization techniques.

In summary, the upside of implicit graph representations is their efficient usage of drawing space for the node primitives. A drawback is that overplotting can hamper perception if nodes are not carefully laid out. Moreover, the directionality of edges may get lost, depending on the chosen layout strategy. Overlap can easily show directionality, as either node u overlaps node v or the other way around. However, adjacency is symmetric and thus requires additional conventions such as drawing from top to bottom or from left to right in order to encode the directionality.

(a) Squarified Treemap.   (b) Information pyramids.   (c) 3D sunburst.

Figure 3.62   Implicit visualizations of a classification hierarchy. Software courtesy of Steffen Hadlak.

Hybrid Representations

Node-link, matrix, and implicit representations are suited for different graph data. Node-link diagrams are good for sparse networks, which have a moderate number of edges. Dense networks with many edges are best visualized using a matrix. Trees, as we just said, are nicely represented by implicit approaches. But what if our graph has sparse and dense parts and includes tree-like sub-structures at the same time?

This is where hybrid representations can help. They combine different representational paradigms. The idea is to mix representations, where possible, in order to utilize the positive aspects and compensate individual disadvantages. Yet, this also adds a new question: Which part of a graph should be represented in which way? Some hybrid representations leave the decision completely to the user, who then has to style the different parts of the graph via interaction. Yet, it is also possible to use methods to detect densely connected, sparse, or tree-like sub-structures in graphs and to assign representational paradigms automatically [AMA07].

A particularly nice example of a hybrid technique is the NodeTrix [HFM07]. As the name suggests, it is a combination of node-link and a matrix representation. Figure 3.63 shows an example of the *Les Misérables* co-occurrence network, where the dense communities are visualized as sub-matrices, whereas connections between these communities are depicted as curved links.

To sum up, node-link, matrix, implicit and hybrid representations are the fundamental options that visualization designers have at their disposal when creating images of graphs. In the next section, we will look at advanced graph visualization approaches taking into account additional data facets.

3.5.3   Visualizing Multi-faceted Graphs

With the visualization approaches described above, we are able to represent the nodes and edges that make up a graph's relational structure. However,

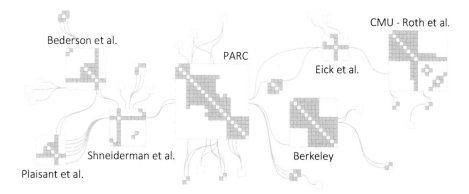

Bederson et al.

PARC

CMU - Roth et al.

Eick et al.

Shneiderman et al.

Plaisant et al.

Berkeley

Figure 3.63 NodeTrix visualization of a co-author network. Adapted with permission by Jean-Daniel Fekete from `www.aviz.fr/Research/Nodetrix`.

we also mentioned additional facets that might be relevant when analyzing graphs. In particular, we named multivariate data attributes, temporal and spatial references, as well as hierarchical groupings of nodes and edges.

This section deals with incorporating these additional facets into the visualization of a graph. Multi-faceted graph visualization is a research topic on its own [HSS15]. The design challenge lies in finding a balanced visual representation of the involved facets that meets the analytic or communicative goals of the user. In order to arrive at a well-balanced representation, it makes sense to follow a two-step design procedure. First, a base representation has to be defined for the primary graph facet whose depiction governs the overall display. Second, the additional facet(s) will be incorporated into the base representation.

For example, if we want to visualize the graph structure together with multivariate attributes, we could choose a node-link diagram as the base representation for the structure. The visual representation of multivariate attributes could be incorporated by varying color, size, or shape of the nodes and the color, width, or dash pattern of the links. Alternatively, if our focus is more on the attributes and less on the structure, we could choose a table-based visual representation for multivariate data elements. The structural aspect could then be incorporated by drawing additional links between the rows of the table.

In general, analogously to what has been said about the combination of views in Section 3.1.2, a visual combination of different data facets can be performed in two different ways. On the one hand, a *temporal composition* can be implemented by utilizing display time to show one facet after the other. On the other hand, the combination of facets can be realized through a *spatial composition*, which utilizes the available display space.

The basic options for spatial composition are juxtaposition, superimposition, and nesting. Figure 3.64 illustrates these options for the case of combining

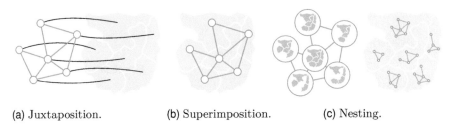

(a) Juxtaposition.   (b) Superimposition.   (c) Nesting.

Figure 3.64   Spatial composition of graph facets in a single representation. Adapted from [HSS15].

structure and geo-spatial context. For juxtaposition, structure and geo-spatial context are shown side-by-side. This results in a balanced view of both facets. However, it is necessary to link the facets via additional visual cues, for example, by arcs as in Figure 3.64a. For superimposition, the facets are drawn on top of each other. In the example in Figure 3.64b, the geo-spatial context is governing the display of the structure in that the geographic regions prescribe the layout of the graph nodes.

This is also the case for nesting, where the nested facet has to obey the positioning and space constraints of the base representation into which we nest. Depending on which facet is nested into which, the visual results can differ significantly. For example, both instances of nesting in Figure 3.64c show the same data, but convey different information. Nesting geo-space within structure (left) tells us which geographic region belongs to a given node, whereas nesting structure within geo-space (right) communicates which nodes belong to a given region.

Let us next look at two concrete solutions that deal with multi-faceted graph visualization. First, we will consider the visualization of dynamic spatial trees as a 3D layered map representation. The second solution visualizes multivariate compound graphs in a multiple-views system.

3D Layered Map Visualization of Dynamic Spatial Trees

As defined earlier, dynamic graphs change over time T, and spatial graphs are connected to spatial references S. Similar to temporal data in general, dynamic graphs are usually analyzed with respect to their evolution over time. Which nodes and edges enter or leave the graph, and which parts remain persistent over time? When dealing with spatial graphs, the analytic interest is focused on the interplay between the graph structure and the spatial context.

Next, we want to demonstrate how tree structures can be visualized together with their temporal and spatial facets. The example we are going to discuss is based on a map display of the spatial frame of reference. The tree structures will be nested into the regions of the map. Superimposition of several map layers is then used to communicate the temporal evolution of the data.

Figure 3.65   Map with tree layouts embedded into selected regions.

Spatial Nesting of Tree Structures   As just said, we start with a display of the spatial facet as a map. The next step is to embed the structural aspect of the trees into the map regions. For the later integration of the temporal facet, it is necessary to use a layout algorithm that assigns fixed positions to the tree nodes. Moreover, the algorithm must be able to adapt the layout to the irregular shape and size of the geographic regions. Both requirements can be fulfilled by combining a point-based layout with a skeleton-based region subdivision [Had+10]. Figure 3.65 shows an example result. We can see how the trees fit nicely into the different map regions.

Temporal Layering and Difference Encoding   In order to visualize time, a separate map layer is created per time step. Following the idea of the space-time cube introduced in Section 3.4.3, the map is tilted into 3D and the layers are stacked along the third display dimension. Figure 3.66 shows three such layers. We can now easily see the structures per time step.

Yet, what can be more difficult to detect are the changes between the individual layers. Therefore, the visualization includes an explicit visual encoding of differences. Red and blue spikes indicate where nodes left or entered the trees. Moreover, orange and brighter blue lines connect those nodes for which the associated data values change significantly. Spikes and lines are always perpendicular to the map layers thanks to the fixed node positions of the tree layout.

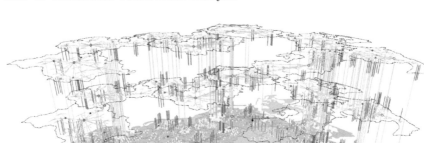

Figure 3.66 Three map layers visualize the data of three consecutive time steps. Spikes and lines indicate differences between the layers.

Note that the static figure cannot reproduce the experience of a live 3D visualization due to the lack of dynamic motion, which is an important depth cue. Because our human perception is very well tuned to interpreting 3D scenes, the actual association of nodes and edges with layers, spikes, and lines can be quite easily interpreted with the running system.

But still, there is a clear limitation of the presented approach: Only a few time steps can be visualized simultaneously. If the number of time steps increases, alternative means must be applied. One option would be to scroll through time, which corresponds to a mapping of the time in the data to the display time.

In general, multi-faceted graph visualization becomes more difficult when the graph itself is large and when the involved facets are extensive as well. In the next section, we illustrate how a multiple-views approach can help in coping with this challenge.

Multi-view Visualization of Multivariate Compound Graphs

As the number of nodes and edges in a graph increases, it becomes more and more difficult to show all of them in a visualization. In such cases, it makes sense to consider groups of nodes and edges as in compound graphs. A compound graph $G = (V, E, H)$ is a graph where a hierarchical grouping is given as a rooted tree H. The set of leaves of H corresponds to the nodes V of G. Non-leaves of H correspond to the groups and are sometimes called meta-nodes or cluster nodes. Groups can be expanded or collapsed in order to study a graph at different levels of abstraction. A multivariate compound graph $G = (V, E, H, A)$ is a graph whose nodes V and edges E have additional data attributes A.

The central objective when analyzing multivariate compound graphs is to understand the relationship of structural properties of the graph and the attributes and groups associated with it. For example, given a subset of nodes being similar in their attributes, do they exhibit similar structural properties? Or, given a certain sub-structure of the graph, do the nodes in that sub-structure exhibit similar attribute values? Or, are the nodes in one group similar to the ones of another group?

To answer these and similar questions, multiple coordinated views can be employed as in the graph visualizations system CGV [TAS09]. From a conceptual perspective, the system solves the problem of multi-faceted graph visualization via juxtaposition of facets, while individual views may work with superimposition.

Multiple Views   Figure 3.67 shows eight different views as provided by the system. The graph structure (V and E) is visualized in the matrix at the left and in the node-link diagram in the center. Selected attributes of A are encoded in these views by varying colors and sizes. A dedicated multivariate representation of the data attributes is offered by the parallel coordinates plot at the bottom. The density of nodes in the graph layout is visualized in a splat view in the bottom-right corner.

What remains to be visualized is the hierarchical organization H of the graph. This is done in the three views at the top. The graphical hierarchy view (top right) shows nodes as colored triangles. This nicely visualizes the sizes of the groups defined by H. The textual tree view (top middle) enables users to read the labels associated with the groups. The 3D tree representation on the left is called Magic Eye View [KS99]. It shows the compound hierarchy

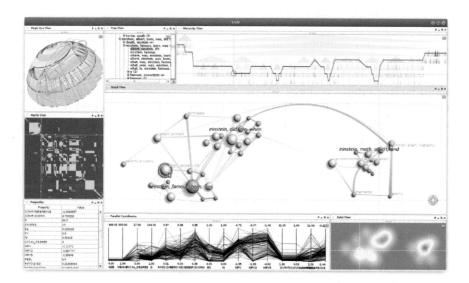

Figure 3.67   Multiple coordinated views for multi-faced graph visualization.

explicitly as red dots and blue links projected onto a hemisphere. Additional orange arcs span the hemisphere to visualize selected links of E in relation to H. Finally, a property view in the bottom-left corner lists the attributes A associated with a selected node or edge. As we can see in this example (and in Figure 1.3 on page 10), there are quite different representations of one and the same graph data. Yet, each view focuses on different facets of the graph.

Coordinated Highlighting In order to make sense of such a multiple-views visualization, it is obviously necessary to mentally connect the different representations. This can be supported by coordinated highlighting, which means that the focused graph element is consistently marked in all views. For example, the node with the label "albert, einstein, life" is marked with a circle in the node-link diagram and with a blue background in the tree view. The parallel coordinates plot visualizes the polyline of the node in red, rather than the regular black. The graphical hierarchy view, the matrix view, and the splat view show crosshairs to mark the node. Changing the focus in either view will change the marks in all views simultaneously.

Additionally, the blue line in the graphical hierarchy view on the top indicates the current cut through the grouping hierarchy. When the user descends or ascends in the hierarchy by expanding or collapsing groups, the blue line will automatically adjust its shape to make clear to the user how deep in the graph the ongoing analysis takes place.

The different views of the graph together with the coordinated highlighting make it possible to study even large multivariate graphs. In general, we can conclude that multi-view visualizations are a good solution when the data are complex in the sense that they contain a diverse set of facets.

With the two examples of multi-faceted graph visualization, the section on how to represent relations among the data comes to an end. We have learned that dedicated methods exist to visualize different classes of graphs. Visualizing multi-faceted graphs requires diverse and rather complex representations and considerable design effort is necessary to create them. In Chapter 5, we will see how graph visualization in general can be advanced further with the help of automatic analytical computations.

3.6 SUMMARY

This chapter explained how data can be mapped to visual representations for interactive visual analysis. At the heart of visualization are two fundamental steps, the visual encoding and the visual arrangement. The visual encoding is about mapping data to graphical marks and visual variables. The visual arrangement describes the layout of the data-representing graphical marks in one or multiple views.

For the most part of this chapter, we described how the fundamental visualization steps are implemented in the context of different classes of data.

In fact, a key message of this chapter is that the characteristics of the data are crucial for designing their visualization. We dealt with several data aspects that need to be considered: Are the data related to time T, are the data embedded in the geographic space S, or do the data contain relations R to be exposed in the visualization? These aspects can also occur in combination. In particular, we discussed the visualization of:

- Multivariate data A

- Temporal data $T \times A$

- Geo-spatial data $S \times A$

- Spatio-temporal data $S \times T \times A$

- Graphs R

- Multi-faceted graphs $R \times T \times S \times A$

As we have seen, the more aspects are involved, the more we can claim that designing expressive and effective visual representations is challenging. And there are still more aspects worth being considered in the context of interactive visual data analysis.

Additional Visualization Aspects   What we have not addressed in this chapter is the aspect of data quality. In fact, most of the data in the wild are "dirty". By dirty we mean there are missing values, there are uncertain values, there are inconsistencies, and there are errors [GS06].

These problems add a whole new layer to the visualization design. For example, the visualization of *uncertain* geo-spatial data not only involves the aspects of space S and data A, but also the aspect of uncertainty U. Considering uncertainty U leads to new questions one may want to investigate. Here is an incomplete list:

- $A \rightarrow U$ – What data are uncertain?

- $T \rightarrow A \times U$ – How do data values and uncertainty evolve over time?

- $T \times S \rightarrow U$ – Where in time and space are the data uncertain?

- $R \rightarrow U$ – Which structural relations are uncertain?

- . . .

Uncertainty is but one aspect among others worth being integrated into the visualization. Another source of information is provenance P. This not only includes data provenance, which explains how the data came about, but also insight provenance, which explains how insights have been generated [Rag+16]. It is obvious that the inclusion of provenance P expands the space of possible analytic questions to an even greater extent.

Prioritized Multi-aspect Visualization   With such a multitude of aspects having an impact on the visualization design, a simple combination of visualization techniques is often not successful. Instead, it is necessary to prioritize certain aspects and adjust the visualization design accordingly [Düb+17].

To this end, scalable visual representations with different levels of sophistication must be developed for each involved aspect. For example, the spatial aspect could be represented in full detail using sophisticated terrain rendering, or as a contour representation that only indicates the main features of the spatial frame of reference. If space plays an important role in an analysis scenario, the sophisticated terrain rendering is employed. If space plays a minor role, the graphically less-expensive contour representation can be used.

Making such importance-driven design decisions for all involved aspects leads to a prioritized multi-aspect visualization. However, this is a topic that requires much more research to arrive at a mature state with a comprehensive consideration of all relevant analysis aspects. Future work in this direction can draw inspiration from a wealth of available visualization techniques, some of which are collected in visualizations surveys.

Visualization Surveys   Since the foundation of visualization as a field in computer science, a variety of visual representations have been developed for all kinds of data. Here, we could only describe the fundamental concepts and procedures and illustrate them with selected examples. Many more techniques are described in the visualization literature.

To help potential users find visual solutions to their data analysis problems, it makes sense to catalogue the available visualization techniques. We already mentioned `browser.timeviz.net` and `treevis.net`, which provide comprehensive lists of techniques for temporal data and tree data, respectively. The InfoVis Wiki curates a compilation of such interactive online surveys at `infovis-wiki.net/wiki/Interactive_Online_Surveys`, including, for instance, surveys on:

- Text visualization: `textvis.lnu.se`

- Biological data visualization: `biovis.lnu.se`

- Financial data visualization: `financevis.net`

- Visualization of scientific literature and patents: `paperviz.org`

- Visualization of dynamic graphs: `dynamicgraphs.fbeck.com`

Often, these online resources are based on books or state-of-the-art reports that study specific visualization issues in greater detail [ML17]. These works should provide the interested reader with plenty of material to delve further into the world of visual data representations.

At this point, we close the chapter on visualization. In the next chapter, we will focus on the role of interaction in the context of visual data analysis.

FURTHER READING

General Literature: [Spe07] • [War12] • [WGK15]

Color Coding: [BRT95] • [HB03] • [ZH16]

Visualization of Temporal Data: [Aig+11] • [Wil11] • [Bac+17]

Visualization of Geo-spatial Data: [Mac95] • [AA06] • [And+13]

Graph Visualization: [vLan+11] • [KPW14] • [Nob+19]

Interacting with Visualizations

CONTENTS

VISUALIZATION techniques provide us with expressive visual representations of data. As humans, we can interpret and make sense of the visual representations and draw conclusions about the underlying phenomena. Yet, if we look only passively at visual representations, we waste much of the potential of visual data analysis. Ideally, we would like to actively engage in a dialog with the data. This includes generating different views, studying specific details, and fine-tuning the visual encoding. All these activities are enabled through *interaction*.

This chapter elaborates on many different forms of interaction for visual data analysis. When discussing interaction, four key aspects are relevant: the human, the tasks, the data, and the technology.

Visual representations are studied and interpreted by *human* users. Based on their impressions, they will interact with the visualization system. The interactions are typically related to analytic *tasks* to be accomplished with respect to the *data* being studied. *Technology* is the mediator in the process of interactive visual analysis. It displays visual output and accepts the user's input. Wrapped into a single sentence, this means: The user solves analytic tasks on data using technology.

This chapter touches upon all of these four aspects. The chapter's first part will focus on the human's role in interactive visual data analysis. In Section 4.1, we motivate the human-in-the-loop approach of interactive visual reasoning and consider conceptual perspectives on humans interacting with visualization systems. The requirements to be fulfilled in order to arrive at useful and usable interaction for visual data analysis will be discussed in Section 4.2. Throughout this chapter, we will keep a keen eye on human aspects.

For the chapter's second part, our attention will shift to the tasks to be carried out interactively and to the data on which they operate. We start with basic low-level interactive operations in Section 4.3. Section 4.4 deals with interactive selection and visual accentuation as fundamental tasks in visual data analysis. In Section 4.5, we continue with multi-scale exploration and navigation in zoomable visualizations. Section 4.6 explains how flexible and light-weight adjustments of visual representations can be made via interactive lenses. As an example of a more complex analytic task, Section 4.7 focuses on supporting visual comparison with naturally inspired interaction. As we discuss interaction for these different tasks, we will also be considering the data aspect. We will see that interaction, very much as the visualization, has to be designed according to the data at hand. This chapter includes interaction techniques for multivariate data, temporal data, spatio-temporal data, and graph data.

The third and last part of this chapter is dedicated to the technology aspect. Section 4.8 goes beyond traditional paradigms and sheds some light on novel ways of interacting with visualizations. As we will see, modern technologies such as touch and tangible interaction or large high-resolution displays open up new possibilities for interactive visual data analysis. Yet, new designs are necessary to make the best of what the new technologies offer.

4.1 HUMAN IN THE LOOP

As indicated, we will begin this chapter with some general thoughts on interaction for visual data analysis and the role of the human within this process. Let us start with the purpose of interaction.

Is interaction necessary at all, can't we let the computer do all the work? Well, if the analytic problem can be formalized precisely and its solution be calculated exactly, then certainly no interaction is needed. Typically, however, data analysis activities are not so simple. Jacques Bertin made this clear even before visualization existed as a field [Ber81]:

"A graphic is not 'drawn' once and for all; it is 'constructed' and reconstructed until it reveals all the relationships constituted by the interplay of the data. The best graphic operations are those carried out by the decision-maker himself."

Bertin, 1981

Bertin's assessment still holds true today. Analytic goals are often ill-defined and exploratory in nature. Sometimes it is even unclear what the desired results are or how they should look like. In Chapter 3, we have seen many different visualization techniques. Deciding for suitable ones and configuring them appropriately require human expertise. It is interaction that enables the user to experiment with different visual encodings and to look at the data from different perspectives.

Interaction is not only helpful for specifying *how* to visualize data, but also for selecting *what* to visualize. Data often capture complex phenomena as an interplay of thousands of multi-faceted pieces of information. The human mind, however, can digest only a limited amount of information at a time. Therefore, the data analysis must be divided into meaningful chunks. It is interaction that enables the user to define such chunks, to navigate between them, and to combine them in order to form a comprehensive understanding [Spe07].

In fact, interaction is not only for *pragmatic* purposes bringing the user closer to a goal. Interaction also serves *epistemic* purposes helping the user form and scaffold a better mental model of the problem being investigated and a better understanding of the tools being used [KM94].

In summary, visualization helps us see things that are otherwise not visible, and interaction allows us to do things that we would otherwise not be able to do. While visual representations may provoke curiosity, interaction provides the means to satisfy it.

4.1.1 Interaction Intents and Action Patterns

Interaction is clearly important for visual data analysis. But what exactly actuates users to interact and what interactions are common? To answer these questions, next we will look into a high-level categorization of interaction intents and a more fine granular list of action patterns.

Interaction Intents

Interaction intents capture *why* users interact with visual representations. Seven broad categories can be identified [Yi+07]. Without going into too much detail, next we will describe the basic idea behind each category.

Mark something as interesting. When users identify something interesting in a visual representation, they typically have the intent to mark it for further investigation. Markings can be *transient* to highlight intermediate findings, or *permanent* to memorize important analysis results over a longer time.

Show me something else. For large and complex data, it is often impossible to visualize all information in a single view. To develop a comprehensive understanding, users have to explore different parts of the data and experiment with different combinations of variables to be shown in the visualization.

Show me a different arrangement. Generating different visual arrangements allows users to study data from various perspectives to obtain different insights. For example, arranging data according to time can reveal trends, while attribute-driven layouts may be better suited to communicate data distribution.

Show me a different representation. The visual encoding is decisive for what can be derived from a visual data representation. Therefore, users want to adapt the visual encoding to suit their needs, be it to carry out different analysis tasks, or to confirm a hypothesis generated from one visual encoding by checking it against an alternative one.

Show me more or less detail. As in real life, users want to look at certain things in detail. On the other hand, users need an overview to keep themselves oriented. During data exploration, the level of detail needs to be adjusted constantly to satisfy the conflicting demands of studying subtleties and seeing the big picture.

Show me something conditionally. It makes sense to restrict the visualization to show only those data that adhere to certain conditions or search criteria that are particularly relevant to the task at hand. Interactively filtering out or attenuating irrelevant data clears the view and allows users to focus on their task.

Show me related things. When an interesting finding has been made, a logical next step is to ask whether similar or related discoveries can be made in other parts of the data. To find, compare, and evaluate such relations, users wish to be connected to them on demand, for example by means of visual links to the related data.

These seven categories cover interactions that are particularly relevant for visual data analysis. Next, we describe two additional categories of more general intents, which are nonetheless quite relevant for visual data analysis as well.

Let me go back to where I've been. Because visual data analysis is an exploratory process, it is usually necessary to try out new views on the data and experiment with alternative what-if scenarios. If exploratory actions do not yield the desired results, users intend to return to a previous state. A history mechanism can keep track of the interaction and allows users to undo and redo operations.

Let me change the interface. In addition to tuning the visual representation to the data and tasks at hand, users also want to adjust the overall visual analysis system. This includes adapting the user interface (e.g., the arrangement of windows or the items in toolbars), and also the general management of system resources (e.g., display resolution and amount of memory to be used).

The presented interaction intents summarize the reasons for an active participation of users in the visual data analysis. The next section focuses on more concrete action patterns.

Action Patterns

An alternative perspective on interaction is offered by action patterns [SP13]. They are more oriented toward *what* users actually do to support the creation of insight during visual data analysis scenarios. Two types of action patterns can be distinguished. An action pattern can be:

- unipolar or
- bipolar.

Unipolar patterns describe human actions that are performed in only one direction. For these actions, there is no natural opposite action. If semantically

TABLE 4.1   Examples of unipolar human action patterns [SP13].

Pattern	Description
Arranging	changes ordering, either spatially or temporally
Assigning	binds features or values to be encoded
Blending	fuses visual representations together to form one entity
Comparing	determines similarities or differences
Drilling	brings out and displays interior, deep information
Filtering	displays subsets obeying certain criteria
Navigating	moves on, through, and around the data
Selecting	focuses on or chooses either individuals or groups

TABLE 4.2   Examples of bipolar human action patterns [SP13].

Pattern	Description
Collapsing/ Expanding	fold in and compact visual items, or oppositely, fold them out or make them more diffuse
Composing/ Decomposing	assemble and join together to create holistic representations, or oppositely, break up into separate components
Linking/ Unlinking	establish relationships or associations, or oppositely, dissociate and disconnect relationships
Storing/ Retrieving	put aside for later use, or oppositely, bring stored items back into usage

meaningful at all, unipolar action can only be reversed by a generic *undo* operation. Selecting, navigating, and comparing are examples of unipolar action patterns.

Bipolar patterns are characterized by the existence of two actions, where one action is the natural opposite of the other. Together, both actions can be used to make progress and to return to previous states. An example of a bipolar pattern is collapsing/expanding. The natural opposite action for collapsing multiple data elements into a single representative is expanding the representative to reveal its individual members.

There are many more unipolar and bipolar action patterns being relevant in the context of interactive visual data analysis. Tables 4.1 and 4.2 list prominent examples. We will return to some of them in particular in dedicated sections. In Section 4.4, we will look at selecting and filtering in more detail. Questions of navigating and drilling will be addressed when we talk about zoomable visualizations in Section 4.5. In Section 4.6, we will blend alternative visual representations by means of interactive lenses. Comparing as a quite interesting action pattern will be discussed in Section 4.7.

At this point it should be clear that interaction is a multi-faceted concept for giving meaning to what is perceived, for collecting relevant information, and for extracting and storing interesting findings. In other words, the human is not only a passive onlooker, but an active participant in a dynamic process. How this process can be modeled and understood conceptually will be discussed next.

4.1.2 The Action Cycle

From a conceptual perspective, the idea of the human-in-the-loop can be nicely explained with Norman's *action cycle*. The action cycle is a general model that describes interaction via several stages of action [Nor88; Nor13]. As shown in Figure 4.1, the human and the system are connected via two phases: an *execution* phase and an *evaluation* phase. The execution phase is related to performing the interaction, whereas the evaluation phase is concerned with interpreting the visual response generated by the system.

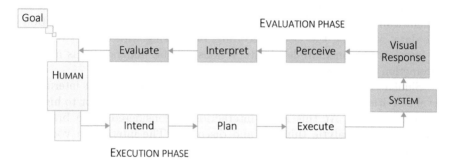

Figure 4.1 Stages of action forming the action cycle. Adapted from [Nor13].

Stages of Action   Figure 4.1 suggests that a motivating *goal* is required before the human engages in some data analysis activities. The action cycle actually starts with *intending* some change that aims at making progress toward the analysis goal. The second stage consists of forming an action *plan* to satisfy the intent. The plan is then *executed* physically in the third stage.

The system will then process the user's request and generate a visual response to be presented to the user. This is where the evaluation phase starts. The new visual representation is *perceived*, and subsequently a conscious *interpretation* takes place. In the final *evaluation* stage, the result of the interaction is compared to the original intent. If there is a mismatch of result and intent, the action cycle has to be re-run. Certainly, the user has to change the course of action to generate an alternative result.

For our explanation, the action cycle ran only once. In practice, though, the action cycle is executed many more times. This is only partly due to discrepancies that may occur between the intent and the result of an interaction. A far more important reason for interaction is the dynamics of data analysis activities. As indicated earlier, interests may vary greatly as humans investigate visual representations of data. Spotting something interesting may raise the intent to study it in detail. Seeing the details may provoke curiosity to look at a different visual encoding. Doing so may lead to new insights or questions that may turn the data analysis in yet another direction creating yet new ideas and objectives.

Levels of Interaction   Depending on the number of runs typically necessary to achieve a goal, one can distinguish different levels of interaction: low-level interaction, intermediate-level interaction, and high-level interaction.

At a low level, interaction is concerned with mapping the fundamental degrees of freedom offered by input devices to basic operations of pointing at and manipulating graphical objects. We may think of such low-level operations as the interaction alphabet or syntax.

At an intermediate level, basic low-level operations are combined to form semantically meaningful data analysis activities. These include navigation in the data, adjusting the visual encoding, or filtering with respect to criteria of interest. Intermediate-level interactions can be understood as a kind of interaction vocabulary.

Similar to how low-level operations are combined to carry out intermediate-level activities, so are the intermediate-level activities a precursor to high-level problem solving. At this level, interaction is considered more broadly as a catalyst for analytic thinking and discovery. The interaction vocabulary is employed to form longer action sequences that support the formation, refinement, and falsification of hypotheses. This includes setting data into relation, extracting high-level features, and organizing the derived knowledge artifacts in an analytic visual-interactive workspace.

As we see, there is a cascade where high-level interaction builds upon intermediate-level interaction, which in turn is based on low-level interaction. At the very heart of this cascade is the action cycle. So, the success of interaction essentially depends on designing the action cycle appropriately. To this end, we need to know more about the requirements for interaction in the context of visual data analysis.

4.2   REQUIREMENTS FOR EFFICIENT INTERACTION

For visual data analysis to be productive, interactions must be designed properly. This section deals with the costs of interaction and discusses requirements that should be met to allow users to accomplish their analysis goals effortlessly.

4.2.1   Interaction Costs

As a matter of fact, the phases of the action cycle occasion costs [Lam08]. This is the reason why they are also denoted as the *gulf of execution* (the intend, plan, and execute stages) and the *gulf of evaluation* (the perceive, interpret, and evaluate stages) [Nor88; Nor13]. The costs can be physical or mental.

Physical Costs are related to performing physical actions, such as moving the fingers or forearms to control the mouse, and to physically sensing the visual response with the eyes. These are largely visceral activities, which happen without the human paying special attention to them.

Physical costs can be significant when engaging in visual data analysis activities. Much of the interaction is of exploratory nature. For such trial-and-error procedures, many repetitive actions can be necessary. For example, functionality may be hidden in cascaded menu structures, which require long and accurate, and hence costly, pointer movements. Exploring the data typically requires constant navigation from one point of interest to another. During the course of an interactive data analysis, pointer mileage and click counts can accumulate and interaction can become a strain.

In a similar vein, perceiving the visual feedback resulting from interaction can be costly. If an interaction leads to much change in the visual representation, the eyes have to sense much information and transmit it to the brain. If visual changes are distributed across the display, the eye muscles will have to work a lot to capture each individual change.

Mental Costs pertain to the stages of intending, planning, interpreting, and evaluating interaction. At these stages, conscious or subconscious activities require the user pay attention to the interaction.

Just like the physical costs, the mental costs can be considerable. Visualization tools are often rich in functionality, and there are various ways to accomplish a task. How does a user know which graphical objects afford what actions upon the visualization? The more options there are, the more difficult it is to identify them, mentally weigh them, and make a decision for one or the other alternative.

Similarly, the visual response can be costly to interpret and evaluate. It is not easy to contrast a refreshed visual representation against how it has looked before the interaction. The difficulty lies in comparing against an image that is no longer being displayed, but instead is only transiently present in the human's short-term memory.

On top of that, visual data representations are usually packed with information, which further complicates the evaluation. If the entire layout of the visual representation changes due to interaction, the user may not be able to follow up and comprehend the effect. On the other hand, the visual response could be hardly visible, because it affects only a few pixels on the screen. This could leave the user wondering if the interaction is ineffective or if the system was not properly notified of the interaction intent.

In the light of interaction costs, we may wonder if we should really try to tackle each and every data analysis problem by means of interaction? No! While interaction is a powerful tool, it is not a silver bullet. Interaction can be a burden when seemingly simple tasks are cumbersome to accomplish due to bad interaction design. Moreover, users may feel uncomfortable with being responsible for even the most basic parameter settings of a visualization. Another problem can be the arbitrariness of visual representations generated through interactive adjustments. It is no longer clear if a feature visible on the screen actually corresponds to a finding in the data or if it is just an artifact caused by an inappropriately set parameter.

This points us to think of interaction in a *less-is-more* way. The system should be responsible for doing its best to relieve the user of unnecessary work. Only as a last resort should input be requested from the user.

From the previous paragraphs, we can conclude that it is essential to narrow the gulf of execution and the gulf of evaluation: Each stage of the action cycle should require as little human effort as possible. Or put differently, interaction has to be designed such that its costs are minimal. In the following, we will see that the directness of interaction is an important factor in this regard.

4.2.2 Directness of Interaction

The directness with which interaction is carried out largely determines how smoothly and efficiently the action cycle can run and how deeply the user can immerse in an interactive dialog with the data.

The importance of what is called *direct manipulation* has been recognized quite early in human-computer interaction research. Hutchins and colleagues advocate direct manipulation and use visual data analysis as their motivating example [HHN85]:

"Are we analyzing data? Then we should be manipulating the data themselves; or if we are designing an analysis of data, we should be manipulating the analytic structures themselves."

Hutchins et al., 1985

Nowadays, direct manipulation is *the* preferred paradigm for interacting with data and their visual representations. The basic idea behind direct manipulation is that the user operates directly on the visual representation of the data using physical actions, and the system provides feedback immediately. According to its classic definition, a direct manipulation interface is characterized by three key properties [SP09]:

- Objects and actions of interest are presented continuously using meaningful visual metaphors.

- The user's requests are expressed through physical actions, rather than complex syntax.

- Actions are rapid, incremental, and reversible, and their effect is immediately visible.

These properties have an immediate consequence for the design of interactive visualizations. A visual representation is no longer only a means to communicate data to the human. Additionally, a visual representation has to provide the means to enable the human to interact with the data and the system in general.

Knowing the basic idea behind direct manipulation, we can next study the question: What does *directness* actually mean?

Separation as an Inverse Measure of Directness

In order to develop an understanding of directness, it is helpful to look at directness from an opposite point of view. In fact, directness is inversely proportional to the degree of separation of the human's action and the systems's response. In other words, we do not achieve directness if there is a high degree of separation. Different types of separation can be detrimental to directness. There are:

- conceptual separation,

- spatial separation, and

- temporal separation.

Conceptual Separation   relates to the different models involved when interacting for visual data analysis. There are the user's mental model, the system's implementation model, and the interface's represented model [CRC07]. The mental model comprises the analytic problem being dealt with and the concept a user has about the system. It abstracts from details and focuses on goal-relevant aspects. Visualization software adheres to an implementation model. This is a formal model full of technical details, algorithmic conventions, and parameterized procedures. The mental model and the implementation model exhibit a large conceptual separation, as illustrated schematically in Figure 4.2. Therefore, there is the third model, the represented model. This model captures what the user can actually see and interact with on the display. The closer the represented model is to the mental model, the more directly can users interact.

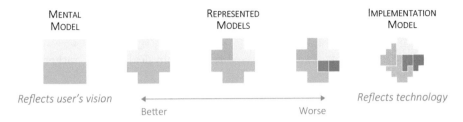

<table>
<tr><td>MENTAL
MODEL</td><td>REPRESENTED
MODELS</td><td>IMPLEMENTATION
MODEL</td></tr>
</table>

Reflects user's vision ← Better ——————————— Worse → *Reflects technology*

Figure 4.2 Conceptual separation across different models. Adapted from [CRC07].

Spatial Separation concerns distances to be covered during the interaction. Large distances can increase interaction costs considerably. Costs accrue when users have to move the pointer across large distances in order to execute certain actions. Spatial separation is also problematic when the eyes have to switch frequently between different parts of the screen when evaluating the system's response. Consider, for example, the scattered data visualization and the corresponding graphical user interface in Figure 4.3. The user interacts with the controls to the right, whereas the visual response becomes visible in the main view. The problem is that action and effect are spatially separated. This can make it more difficult to understand the action-effect causality. The user may have to shift the attention back and forth between the user interface and the main view several times to comprehend how certain parameters influence the visual representation of the data.

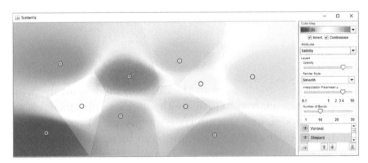

Figure 4.3 Spatial separation between the graphical user interface (right) and the visual representation in the main view (center).

Temporal Separation is about the latency between the user's action and the system's visual response. Ideally, a response should be provided within 50–100ms [Shn94; Spe07]. As a matter of fact, if the latency is too high, the efficiency of interactive visual exploration degrades [LH14]. However, the computations involved in processing the visualization transformation and generating the visual feedback can take a considerable amount of time. For example, the visualization in Figure 4.3 consists of scattered data points, their corresponding Voronoi diagram, and a smooth coloring in the background. With respect to latency, the individual data points can be rendered virtually at no time cost. Even the Voronoi diagram can be computed quite quickly given the fact that there are only ten data points to be processed by a $\Theta(n \log n)$ algorithm. For the smooth coloring, however, a color value needs to be computed for every pixel of the main window. Already for a moderately sized window with a $1{,}280 \times 1{,}024$ resolution, $1{,}310{,}720$ pixels need to be colored. On a 4K display, it will be $8{,}294{,}400$ pixels. This can cause a noticeable delay leading to the adverse effects of efficiency degradation.

All three types of separation impair directness and hinder smooth and efficient visual data analysis. Therefore, reducing separation should be a top priority throughout the visualization development process. This concerns both the design and the implementation. Interaction and visual feedback should be designed so as to minimize conceptual and spatial separation. Implementation-wise, algorithmic efficiency is important to reduce temporal separation.

Scenarios of Different Directness

We previously discussed directness on a theoretical level. But how does directness manifest in practice? Next, we sketch five scenarios illustrating different degrees of directness (or separation) when interacting with a visualization. For these scenarios, we assume a user has spotted an interesting group of nodes in a graph visualization and wants to zoom in to look at them in detail. There are several alternatives to accomplish this by interacting with the system.

1. **Source Code Editing** On the implementation level, a visual representation is defined by source code. Only a few lines of code need to be edited to set the visualization view to where the nodes of interest are located. The altered code is compiled and run to see if the visual result is as expected. If not, the procedure is re-run until the user is satisfied. Changing code lines, re-compiling, and test-running the visualization is the least direct form of interaction, as it exhibits large conceptual, spatial, and temporal separation.

2. **Scripting Commands** Alternatively, the visualization may offer a scripting interface allowing the user to enter commands to zoom the view. Once issued, the commands take effect immediately while the visualization is running. In this scenario, no separate compilation is necessary, which reduces the temporal separation. But still the interaction is rather indirect, and several commands may be necessary before the view fits as desired.

3. **Graphical Interface** The field of view is displayed in a graphical interface alongside the visualization. Standard controls such as buttons and sliders

allow the user to easily shift the view and control its zoom factor. Any changes are immediately reflected in the graphical interface and the visualization. Given the fact that the graphical interface represent the view status and at the same time serves to manipulate it, the conceptual gap is narrowed. Yet, the interaction (with the controls) and the visual feedback (in the graph visualization) are still spatially separated.

4. **Direct Manipulation** The user zooms the view directly by drawing an elastic rectangle around the nodes to be inspected in detail. This is a rather simple press-drag-release operation when using the mouse. During the interaction, visual feedback constantly indicates the frame that will make up the new view once the mouse button is released. Any necessary fine-tuning can be done using the mouse wheel. In this scenario, the manipulation of the view takes place directly in the visualization. There is no longer a spatial separation between the interaction and the visual feedback. Or is there?

5. **Direct Touch** Indeed there remains some degree of separation. The interaction is carried out with the mouse, whereas the feedback is shown on the screen. To obtain a yet higher degree of directness, the interaction can alternatively be carried out using touch input on the display. The basic principle of sketching an elastic rectangle to specify a new view is maintained, but the action is performed using a finger directly on the display. Now, the interaction takes place exactly where the visual feedback is shown. A truly direct way of zooming in on a node-link diagram.

These five scenarios illustrate different degrees of directness when interacting with a visualization. Certainly, our list is not exhaustive, but it does contain some general ideas on how interaction can work in practice.

From the examples, we may be inclined to think that direct touch is the best solution. But that is not quite true, we have to differentiate. Each scenario involves its own benefits and drawbacks. Direct touch, for example, tightly links interaction and visual feedback. However, the hand being used for the touch interaction may occlude the user's view on important information. Scripting commands, for another example, can be difficult to specify for novice or casual users. Yet, expert users may be faster and more precise issuing commands than using a graphical form of specification. Moreover, commands are easily reproducible. Even the least direct scenario of modifying lines of source code has some benefits. It enables a visualization developer to test-drive several alternative visual representations without the need to set up and implement dedicated interaction mechanisms. This can be helpful when rapidly prototyping visualization software.

From our discussion we can see that it is important to carefully weigh the means of interaction that a visualization should exhibit. It can very well be that several alternatives are necessary to be able to adapt the interaction to different user audiences.

In any case, we need to design the interaction such that it is useful and usable. The next section provides some guidelines that can help us achieve this goal.

4.2.3 Design Guidelines

Usability and *user experience* are important factors for the development of usable and useful interaction. They subsume several objective and subjective quality criteria for interaction, including efficiency, predictability, consistency, customizability, satisfaction, engagement, responsiveness, and task conformance, to name only a few.

General Rules

Guidelines can inform the design of efficient interaction. Here, we list the golden rules by Shneiderman and Plaisant [SP09]. They are not specifically related to visual data analysis, but still offer valuable advice.

1. **Strive for consistency.** Consistent actions should be required in similar contexts and be responded to consistently.

2. **Cater to universal usability.** The system should be usable for novices, casual users, and experts alike.

3. **Offer informative feedback.** For each possible action, there should be informative feedback appropriate to its importance.

4. **Design dialogs to yield closure.** Action sequences should have a well-defined beginning, middle, and end.

5. **Prevent errors.** The system should prevent serious errors and be able to recover from minor problems.

6. **Permit easy reversal of actions.** Actions should be reversible to allow for undoing accidental actions and to encourage exploration.

7. **Support internal locus of control.** Users should be given the feeling that they are in charge, not the computer.

8. **Reduce short-term memory load.** Short-term memory load should be limited to seven plus minus two chunks of information.

Fluid Interaction

In the context of visual data analysis, one can operationalize the aforementioned general rules to define what is called *fluid interaction* [Elm+11]. The idea of fluid interaction includes three guiding principles:

- Promote flow

- Support direct manipulation

- Minimize the gulfs of execution and evaluation

We already talked about direct manipulation and minimizing the costs of interaction. The first principle, promote flow, though is worth a few words of explanation. By "promoting flow" it is meant that interaction should be designed so that users can totally immerse in the data analysis activities. This requires balancing the difficulty of analytic tasks and the skills of the user, giving the user a sense of control over what is going on, having the right tools available at the right time, and providing an overall rewarding experience.

Heer and Shneiderman underline the significance of the fluent, direct, and human-centered character of interaction in visualization [HS12]:

"To be most effective, visual analytics tools must support the fluent and flexible use of visualizations at rates resonant with the pace of human thought."

Heer and Shneiderman, 2012

With this statement, we end the first part of this chapter, in which we considered interaction mostly at a conceptual level, focusing on the role of the human-in-the-loop. The upcoming second part will be more concrete in terms of interaction tasks performed when analyzing different data. Based on a brief look at low-level interaction in the next section, later sections will introduce specific interaction concepts for interactive visual data analysis.

4.3 BASIC OPERATIONS FOR INTERACTION

We already mentioned that visual analysis involves high-level interaction and builds upon intermediate-level interaction techniques, which in turn are based on low-level operations. These basic operations are the subject of this section, which is intended to bridge our previous conceptual considerations and the interaction techniques described in the sections to come.

As visual data analysis is largely based on interacting with visual representations, we will focus on fundamentals of *graphical interaction*, that is, the procedure of performing physical actions in the real world that change virtual graphical objects on a computer display.

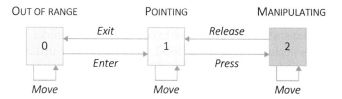

Figure 4.4 Three-state model of graphical input. Adapted from [Bux90].

4.3.1 Taking Action

When interacting with visual representations, there are two basic questions. We have to define, first, *where* our interaction should take effect and, second, *what* the effect should be. Consequently, we distinguish two basic operations:

- point and

- manipulate.

Pointing allows us to define which graphical objects we want to interact with and the *manipulation* defines what should happen with the objects. For example, we may point at a node in a node-link-diagram to mark it as relevant, or we point at a slider handle and then manipulate it to adjust a filter threshold.

Physically, pointing and manipulation can be carried out in different ways. The necessary actions are typically related to movements that are tracked with an interaction device. Moving our hand to control a computer mouse is a prominent example. Moving our fingers allows us to press down a button or rotate the mouse wheel. We are also used to moving fingers and hand across our mobile phone's surface.

Conceptually, graphical interaction can be described with a three-state model [Bux90]. Figure 4.4 shows the states as boxes and the transitions between them as arrows. Consider state 1 first. This is the state we are in when moving the mouse to point at something on the screen. As soon as we press a button, we transition from state 1 to state 2. This is how manipulations are triggered. For example, we can move the mouse to drag a filter slider. Upon releasing the button, we will return to state 1. If the mouse is out of reach, for example, if we move it beyond our application window or lift it from the table, we fall back to state 0, in which movements of the input device do not have an effect.

Note that the previous paragraph describes only one simplified instance of the three-state model. Different configurations exist for different input modalities. For example, touch interaction does not support state 1 out of the box. Once our finger is on the surface, we immediately begin manipulating what is under our fingertip. Clever interaction design and additional timing information are necessary to compensate for the missing state. On the other hand, a mouse with multiple buttons will lead to additional states and transitions. These can be exploited to provide more ways of interaction, but also complicate design and use.

Modes of Interaction

With the three-state model, we now have a conceptual understanding of how interaction is performed on a low level. We can further utilize the model to distinguish two characteristic modes of interaction. There are:

- discrete (or stepped) interaction and

- continuous interaction.

Triggering a short state change can be considered a *discrete interaction*. Examples would be pressing and immediately releasing the mouse button or briefly tapping on a touch-enabled surface. In both cases, we temporarily enter state 2 to trigger a manipulation. We may have clicked a button to change the visual encoding or tapped a data element to mark it. Discrete interaction is most useful for occasionally making a selection when there are only few alternatives to choose from.

However, in visualization scenarios, we often face very many alternatives. This not only concerns the many data elements that we may want to study in detail. There are also many parameters with potentially large value ranges to control the visualization transformation. Exploring the data and browsing different parameterization in a purely discrete fashion would certainly be cumbersome.

This is where *continuous interaction* comes into play. For continuous interaction, we remain in a manipulating state for a longer time while continuously moving the input device. For example, when adjusting a filter slider, each incremental cursor movement results in a separate visual response. The advantage is that a larger space of options can be scanned in a short period of time using a single continuous gesture. This makes continuous interaction particularly useful for exploratory visual data analysis, where testing many *what-if* alternatives is a common task.

4.3.2 Generating Feedback

So far, we have considered the fundamental low-level actions that users may perform. Yet, to complete the action cycle, a user request always requires a visual response by the system. From a system perspective, two operations need to be carried out:

- update and

- refresh.

The *update* operation is concerned with changing the internal state of the visualization based on the action performed by the user. The *refresh* operation, on the other hand, is responsible for presenting a new visual representation that reflects the internal change.

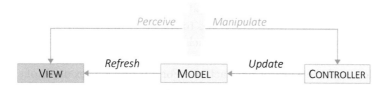

Figure 4.5   Model-view-controller pattern.

Again, a conceptual point of view can help us clarify the update and refresh operations. This time, we refer to the model-view-controller (MVC) pattern [KP88]. MVC is a software design pattern for graphical user interfaces. Figure 4.5 illustrates a variant simplified for the context of this text. In our case, the model consists of the data and all parameters of the visualization transformation. The views correspond to visual representations of the data and auxiliary information. The controllers conceptually capture all different ways of interaction.

As we see from the figure, once the user performs some manipulation, the controller sends an update request to the model. This update can be as simple as a mode switch, but may also be as complex as performing some analytical processing of the entire data. Once the model has updated its internal state, the views are notified to refresh themselves to reflect the new model state. Again this could require a simple repaint using different colors, but also a complete rebuild of the visual data representation.

In terms of temporal separation of action and response as discussed earlier in Section 4.2.2, both update and refresh are critical. Ideally, they run so fast that the user has the feeling of having a system that reacts immediately.

Types of Visual Feedback

Update and refresh are the operations that generate the visual feedback. But how can the feedback itself be characterized? Analog to the modes of interaction, we can distinguish two types of visual feedback:

- static feedback and

- animated feedback.

For *static feedback*, the system creates a single new visual representation to replace the old one. The new image immediately corresponds to the system's new state. Such an abrupt visual change can be useful to draw the user's attention toward the significant differences caused by the interaction. However, there is also a drawback: With static feedback, it might be difficult to comprehend how the new visual representation has evolved from the old one. For example, when switching to an alternative graph layout algorithm, the graph nodes may be located at completely different positions. In such situations, users will have a hard time maintaining their mental model of the data.

With *animated feedback*, the visual representation changes gradually. The system generates a series of responses to smoothly interpolate the display from its current state to the new one. Each intermediate step represents only an incremental update, which makes it easier for users to follow the changes and to understand the effect of their actions. For our previous example of switching the graph layout, nodes can be moved smoothly from their original locations to the new ones. However, a difficulty with animated feedback is that it takes time to complete and hence delays the action cycle. Therefore, animated visual feedback has to be designed with care, but it is worth the effort.

To summarize, pointing and manipulation are the basic operations to be carried out by the user. As we have learned, interactions can be discrete or continuous, where the latter is particularly useful in the context of visualization. Updating the data model and refreshing the visualization views are the basic operations to be carried out by the system in order to create visual feedback. The feedback can be static or animated, where the latter can help users better understand the effect of their actions. Together, the basic operations as described in this section are the fundamental building blocks for interactive graphical systems.

In the following sections, we add the necessary semantics to create useful interaction techniques for visual data analysis. Next, in Section 4.4, we start with interactive selection as a preparatory step for many other operations such as highlighting, filtering, or even modifying data. How data can be explored interactively at multiple scales will be explained later in Section 4.5.

4.4 INTERACTIVE SELECTION AND ACCENTUATION

From our regular work with computers, we know that selecting something on the display is a fundamental and frequently used action pattern. In the context of visual data analysis, interactive selection takes a similarly central role. In fact, selection is a door opener for visual data analysis. It allows us to divide the analysis into smaller manageable subtasks simply by marking parts of the data to be relevant to the question at hand.

Conceptually, selection can be described as follows. Assume we want to analyze a visual representation of some dataset $V(D)$ that contains a lot of information. As we cannot digest all information at once, we successively concentrate on different subsets of the data. Let $D^+ \subset D$ denote what we are currently interested in. The data $D^- = D \setminus D^+$ are not in our focus, and hence, are considered less relevant for the time being. D^+ and D^- are constantly in flux as the user's attention and interests change during the data analysis.

The distinction of D^+ and D^- implies two things. First, interactive visual analysis requires means to enable users to specify what is relevant to them. This aspect will be discussed next in Section 4.4.1. Second, the visual representation of the data must be adapted in such a way that the relevant data stand out. Corresponding accentuation strategies will be introduced later in Section 4.4.2.

4.4.1 Specifying Selections

There are two points of departure for specifying a selection: Data can be selected based on their location on the display and based on the actual data values. In the former case, the selection takes place directly in the visualization. This is also known as *brushing* [BC87; MW95]. In the latter case, selection criteria are defined on the data-level using dedicated controls, which is also referred to as *dynamic querying* [AS94; Shn94]. Next, we take a closer look at the two alternatives.

Interactive Brushing

Interactive brushing works by marking a part of the visual representation and the data shown in that part get selected. With interactive brushing it is possible to specify three categories of selections: point selection, range selection, and composite selection.

Point Selection   A very basic form of brushing is to point at a location of interest followed by a discrete interaction, such as a mouse click or a tap on the screen. But how do we know which data reside at the selected point?

To answer this question, the visualization transformation has to be inverted. This is also called *picking*. While the visualization transformation maps data to graphics, picking reverses this mapping in order to get from graphics to data. Depending on the underlying graphics, different methods can be employed for picking, including lookup-buffers, geometrical tests, or plain mathematical calculations.

Range Selection   There are two classic techniques for marking entire 2D ranges in a visualization: *rubberband selection* and *lasso selection*. Both are continuous interactions that employ a press-drag-release gesture for defining the selection. The rubberband selection uses an elastic shape, for example a rectangle, which makes it easy to mark orthogonal parts of the visualization. If the layout of the data is rather irregular, a free-form lasso selection is more practical. It offers more flexibility, but comes at higher interaction costs, because more (and more accurate) physical movements of the pointer are necessary.

Rubberband and lasso define a geometric shape (rectangle or free-form shape) based on which the actual selection is made. There are two alternatives for that. For data to be selected, they must either be *included in* or *intersect with* the shape. Inclusion and intersection have different characteristics in terms of accuracy and effort. Which alternative is more suitable depends on the graphical objects to be selected. For smaller regular objects like the nodes in the graph layout in Figure 4.6a, inclusion is practical. We can easily and quite accurately include them in a rectangle or lasso. In contrast, larger irregular objects such as the geographic regions of the map in Figure 4.6b

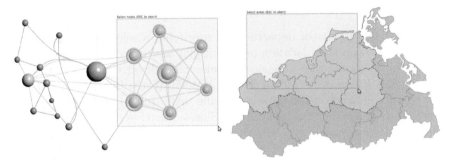

(a) Selection by inclusion.   (b) Selection by intersection.

Figure 4.6   Rubberband selection for marking multiple data elements.

are difficult to mark without adding too many unwanted objects. In this case, an intersection-based mechanism is more practical because it is easier to just *touch* the data we want to mark. Enclosing the data entirely is not necessary.

With the techniques described so far, we can select the data from one particular point or area. But how can we select data from multiple origins? Marking them with a lasso selection is cumbersome at best, if not impossible. Therefore, we need interactions that enable us to combine and edit selections.

Composite Selection   There are various ways to compose multiple select operations. In fact, with 524,288, the number of theoretically possible combinations of add, remove, subtract, intersect, union operations on selections is extremely high [Wil96]. Therefore, it makes sense to follow established standards.

Multiple selections can typically be composed using modifier keys. Holding down the CONTROL key will usually toggle the data's selection state. That is, unselected data become selected and vice versa. When data are arranged in a linear order, as for example top to bottom for the trajectories in Figure 4.7, then the CONTROL and SHIFT keys can be used in combination to select multiple subranges.

Similar variations of the selection behavior can be designed for area selections via rubberband or lasso. The SHIFT and CONTROL keys can be used to add and remove data to and from a selection, respectively. Holding both keys simultaneously enables the creation of intersections.

In summary, interactive brushing together with the options for composing selections enables us to specify our interest directly in the visualization. The advantage is that we can easily select what we see. On the other hand, the selection is based solely on the data's visual arrangement. This can lead to situations were selections are unnecessarily complex to define. Therefore, it makes sense to complement interactive brushing with dynamic queries.

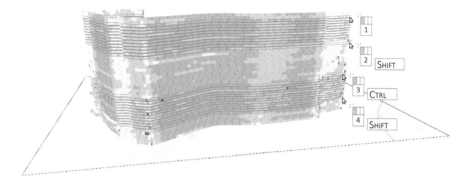

Figure 4.7 Four steps of selecting multiple trajectories using modifier keys.

Dynamic Queries

Data-based selection via dynamic queries is helpful when visually based selection is impractical. Consider, for example, the trajectory visualization in Figure 4.8a and suppose we are interested in selecting the low-speed segments in red and yellow. Brushing them is hardly possible. Rubberband selection is infeasible due to the irregular arrangement of the segments, and lasso selection would require a costly circumvention of the high-speed segments. Composite selection would be possible, but it would take very many individual operations.

So why not select low-speed segments based on the speed values directly? To this end, we need a dedicated representation of the value range of the speed attribute and a mechanism that allows us to mark what we are interested in.

Interactive Legends Interactive legends are a helpful tool for specifying dynamic queries [RLP10]. The example legend shown in Figure 4.8b consists of multiple sections, each is associated with a specific data interval. Users can simply click on a section to (un)select all trajectory segments that fall into the corresponding data interval. In our example, the segments with high speeds have been unselected to concentrate on the data where movements are slow. This would have been very hard to accomplish with interactive brushing, whereas querying via the legend took but a few clicks.

Query Sliders Dynamic queries can also be done in a continuous fashion. For an illustration, let's assume we want to select the red high-degree nodes from the graph visualization in Figure 4.9a. As we can see, the graph layout does not correspond to the characteristics based on which we want to select, which rules out interactive brushing as a suitable selection method. Moreover, we are not quite sure about what 'high-degree' means precisely and need a bit of flexibility to experiment with potential thresholds.

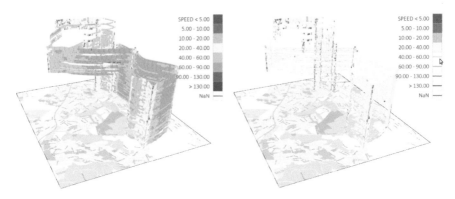

(a) How to select slow-speed segments?   (b) Select via an interactive legend!

Figure 4.8   Selecting segments based on their color, which represents speed.

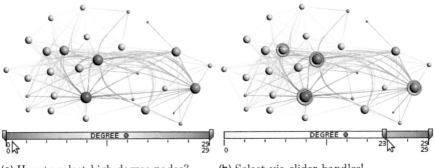

(a) How to select high-degree nodes?   (b) Select via slider handles!

Figure 4.9   Selecting nodes based on their data attributes.

This is where query sliders enter the stage. They are commonly applied in visual analysis scenarios [Eic94]. The exemplars in Figure 4.9 consist of a scale representing the value range of the node degree and two handles enabling the user to specify, in a continuous fashion, a data interval of interest. In Figure 4.9b, we have adjusted the slider so as to select all nodes with a degree between 23 and 29.

The advantage of dynamic query sliders is obvious: We can precisely define our interest based on the characteristics of the data. Observing the visual representation while specifying the selection further enables us to evaluate on the fly if the result suits the task at hand. Adding more handles allows for selecting multiple distinct data ranges. It is even possible to combine multiple sliders to create sophisticated multi-criteria selections.

However, a disadvantage of dynamic queries is that the interaction no longer takes place directly in the visual representation of the data, but with a dedicated interface, a legend or a slider. This may lead to spatial and conceptual separation. Therefore, it is important to strive for a tight integration of the selection interface and the visualization. In our previous examples, color has been used as the link that connects interface and visualization. This makes it easier to comprehend how a selection at the interface will affect the visual representation.

The interactive mechanisms described above enable users to specify their interest in the data. The next step is to adapt the visualization such that relevant data stand out, allowing the user to fully concentrate on them. We have already seen several examples of how this can be done in the figures of this section. Next, we will systematically discuss different strategies of emphasizing or attenuating certain elements of a visualization.

4.4.2 Visual Emphasis and Attenuation

Given a visual representation of the data $V(D)$ and a selection of relevant data D^+, the question that we deal with now is how to visually distinguish $D+$ from the rest of the data D^-? What is needed are additional encodings that either visually emphasize V^+ or visually attenuate V^- certain parts of the data. An alternative is to suppress parts of the data altogether.

Using D^+, D, D^- and V^+, V, V^- as a conceptual basis, one can implement different strategies:

Highlighting Highlighting emphasizes the data of interest $V^+(D^+)$ in contrast to a regular visualization of less-relevant data $V(D^-)$.

Dimming Through dimming, we can attenuate less-relevant data $V^-(D^-)$, while data of interest are visualized regularly $V(D^+)$.

Filtering For filtering, only the relevant data are visualized regularly $V(D^+)$, whereas all other data are omitted.

A decision for either strategy should be made carefully. As a general rule of thumb, the visual feedback should be effective, but with only minimal side effect on the regular visualization.

There are two important influencing factors: the frequency of change of the selection and the size of the selection. Highlighting and dimming contrast the regular visualization V against an alternative encoding (either V^+ or V^-). If D^+ can be assumed to be small, as for example when hovering the cursor over individual elements of a visualization, then highlighting is a sensible choice. If D^- is small, as for example when we want to get rid of a few outliers, then dimming makes sense. Filtering effectively clears the view and allows users to fully focus on D^+. A disadvantage, though, is that the user is no longer aware of the presence of D^-. Therefore, filtering is typically applied only when a selection is stable for a certain subtask during the visual analysis.

Theoretically, further strategies are possible. For example, one could redundantly emphasize relevant data and dim or filter less-relevant data

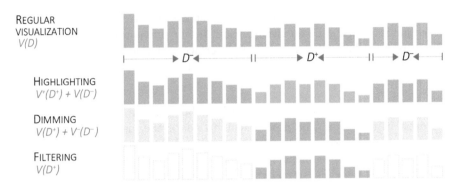

REGULAR
VISUALIZATION
$V(D)$

HIGHLIGHTING
$V^+(D^+) + V(D^-)$

DIMMING
$V(D^+) + V^-(D^-)$

FILTERING
$V(D^+)$

Figure 4.10 Strategies for visual emphasis of relevant data and attenuation of less-relevant data.

$(V^+(D^+) + V^-(D^-)$ or $V^+(D^+)$, respectively). While this would generate a stronger contrast, the regular encoding V would not be used anymore. As this violates the rule of thumb not to interfere too much with the visualization, such redundant encodings are hardly applied in practice.

Let us next illustrate our conceptual considerations with the example in Figure 4.10. Our starting point is a simple visual encoding that shows a time series as bars of varying size in neutral gray, our V. In order to emphasize and attenuate selected parts of the data, we vary color in this example. For V^+, we use a dedicated highlighting color that can be easily distinguished from the neutral gray. Dimming is accomplished by using a lighter gray for V^-, which blends well with the white background. Note that filtering is indicated via dashed outlines, even though the data are actually invisible.

Figure 4.10 shows but a simplified example to illustrate how selected data can be accentuated. In real-world applications, finding suitable visual encodings is considerably more difficult. Usually, the regular encoding V already occupies much of the visual resources (e.g., color, size, position) for the purpose of effective and efficient visualization of the data. The difficulty is to come up with adequate V^+ and V^- to achieve emphasis and attenuation.

The actual design challenge is that V^+ and V^- should not (or only minimally) interfere with V, but need to be sufficiently expressive to contrast D^+ against D^-. This design challenge has to be solved depending on application needs. A general strategy though is to use visual variables that are not yet used for the regular visualization. If this is impractical or impossible, it can even be necessary to embed additional graphical elements, such as outlines or halos into the visual representation. Of course this has to be done sparingly to avoid cluttering the visual representation of the data.

Figure 4.11 provides a few more practical examples for visual emphasis and attenuation for graph visualization. Figure 4.11a shows a node-link diagram where node size and color encode certain node attributes. We want to concen-

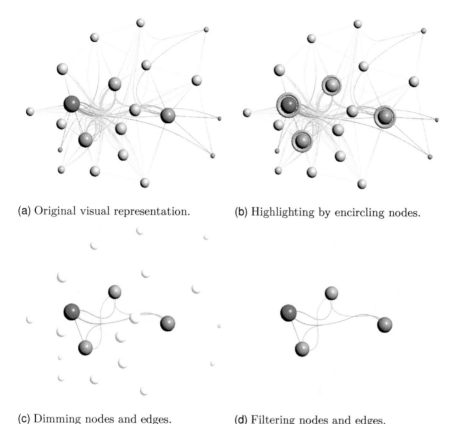

(a) Original visual representation.

(b) Highlighting by encircling nodes.

(c) Dimming nodes and edges.

(d) Filtering nodes and edges.

Figure 4.11 Visual feedback for selections in visual representations of graphs.

trate on the four larger red-orange nodes. It is not possible to highlight the nodes using a special color, because the nodes are already color-coded. Instead, Figure 4.11b shows the selected nodes highlighted via additional graphical primitives, circles in this case. While the nodes are now clearly marked as relevant, our attention may still be distracted by the presence of too many nodes and edges.

Therefore, let us further draw the user's attention to the relevant information by dimming all unselected nodes and their incident edges. In Figure 4.11c, the data of interest stand out clearly now, and it is easier to see how they are interconnected. For an even more focused view, we can filter all irrelevant information as in Figure 4.11d. While this removes any distraction, it also obliterates awareness of the filtered information.

Looking at the figures, one can easily imagine alternative solutions, and indeed the design space for visual emphasis and attenuation is large [Hal+16]. Any visual variable can theoretically be used as a visual cue, including color, size, position, texture, blur, and even pulsing animations or blinking [WH04]. Particularly promising are visual cues that are perceived pre-attentively, which means they immediately draw our attention [HE12].

Deciding for either of these options is not an easy task, even for the binary distinction of D^+ vs. D^- discussed so far. Matters quickly become complicated when it comes to encoding multiple selections simultaneously in a single visualization image.

In conclusion, the discussion in this section makes clear that expressive visual accentuation is as important as effective means for interactively specifying selections.

4.4.3 Enhanced Selection Support

So far, we have discussed basic means of interactive selection. In the following, we will introduce methods to enhance the work with selections in visual analysis scenarios. *Smooth brushing* will allow us to go beyond binary selections. *Brushing & linking* helps us propagate selections across multiple visualization views. We will further see how automatic methods can be used to reduce the costs of interactive selection.

Smooth Brushing

Up to now, our selections define a binary distinction of data into selected and unselected. Smooth brushing is a concept that breaks the barrier of discrete binary selection [MW95; DH02a]. The idea is to assign to each data point a continuous *selectedness* from the interval [0..1]. The extremal values 0 and 1 stand for unselected and selected, respectively. Data points with a selectedness greater than 0 and smaller than 1 are *somewhat* selected. In other words, we are working with a kind of fuzzy selection.

Obviously, a fuzzy selection requires dedicated interaction and visual feedback. Let us illustrate this with the parallel coordinates plots in Figure 4.12. As with regular brushing, the user marks the data range to be assigned a selectedness of 1. This is shown in Figure 4.12a. For smooth brushing, the selectedness automatically spreads beyond the initially defined range until it gradually fades out to zero, as shown in Figure 4.12b. The automatic gradual spread is key to capturing features that by nature are not crisp and clear, and hence, costly to define.

In order to provide visual feedback for smooth selections, it makes sense to consider selectedness as an additional attribute to be visualized alongside the data. The binary selection in Figure 4.12a shows selected and unselected data tuples as black and gray lines, respectively. The smooth selection in Figure 4.12b encodes selectedness by varying the lightness of the gray tones. This allows us to see even subtle differences in selectedness.

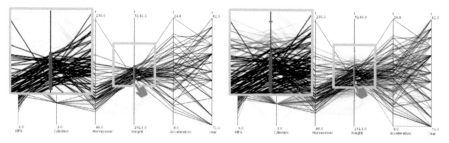

(a) Binary selection in the range.   (b) Fuzzy selection beyond the range.

Figure 4.12   Brushing a range (red) of an axis for binary and fuzzy selection.

Brushing & Linking

Regardless of whether we use regular or smooth brushing, it is always executed on a *single* visual representation. However, visual data analysis often requires working with *multiple* representations. The question that arises is how selections can be propagated across several visualization views?

Consider for example the graph visualization system in Figure 4.13. It consists of four views: a textual tree view to the left, a hierarchy view on the top, a node-link view in the center, and a parallel coordinates view at the bottom. Each view allows us to select data that interests us. But how can the selection be kept consistent across all views? Manually replicating the selection in the other views is certainly impractical. What is needed is a concept to automatically link all views. Or, as Buja and colleagues put it [Buj+91]:

"Multiple views, however, should not be regarded in isolation. They need to be linked so that the information contained in individual views can be integrated into a coherent image of the data as a whole."

Buja et al., 1991

Brushing & linking (or focusing and linking) provide the answer to the aforementioned questions [BC87; Buj+91]. Brushing is what the user does manually. The linking is done automatically by the system. The basis for the linking is a dedicated selection model where views are registered. As all views share the same model, a selection made in one view is automatically available in all other views. The only responsibility of the individual views is to represent the selection visually. For the example in Figure 4.13, the tree view shows a blue label background for selected nodes, the hierarchy view uses dark outlines, the node-link view encircles selected nodes, and the parallel coordinates view dims data that are not selected.

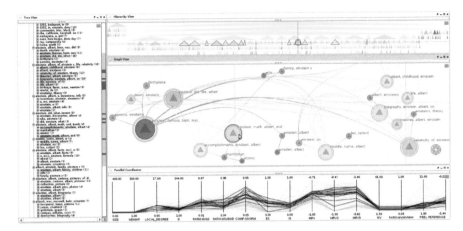

Figure 4.13   Brushing & linking in a multiple-views graph visualization.

So, with brushing & linking, we only need to perform the selection in one view instead of several. Moreover, we can revise the selection in any view, no matter where the selection has been triggered, and the selection consistency is maintained automatically. This gives us much flexibility in specifying our interest in selected parts of the data.

Automatic Selection Support

Interactive selection can particularly benefit from integrating automatic methods that take over otherwise manual selection steps. The general idea is that only an initial mark has to be placed by the user. From there, an automatic procedure completes the selection.

We have already seen instances of this idea in the previous sections. For smooth brushing, selectedness is *automatically* spread beyond the initially brushed data interval. For brushing & linking, selections are *automatically* propagated across multiple visualization views.

Developing dedicated selection algorithms is sensible if manual selections are very costly. This is the case, for example, when selecting clusters in 3D point clouds. Purely interactive selection in 3D point clouds is notoriously tedious and time-consuming work. Automatic selection procedures can facilitate this work [Yu+12]. Based on a simple 2D lasso selection, an algorithm automatically expands the selection along analytically determined structures in the 3D point cloud. This effectively relieves the user of many incremental and costly 3D selection steps.

The idea of automatic selection support can be generalized in various ways [HAW08]. For this purpose, it is common to transform graphical selections into abstract models that describe selection rules or constraints rather than the selection state of individual data elements [Che04]. Such declarative models help us apply the same selection to different datasets, maintain selections even when the data vary over time, or derive relaxed selections when broadening the view on the data is necessary.

With this outlook on automatic support, we end the section on interactive selection and accentuation. We have learned that being able to mark and emphasize data according to *varying interest* is of fundamental importance for visual data analysis. In the next section, we will see that it is just as important to be able to examine data at *varying scales*, especially when dealing with large amounts of data.

4.5 NAVIGATING ZOOMABLE VISUALIZATIONS

Data are usually explored, because it is not clear upfront where interesting patterns are located in the data. Interactive selection as described in the previous section enables us to investigate *different parts* of the data. Yet, this is only one part of the data-exploration equation.

A second reason for data exploration is that it is in many cases also unclear how deeply valuable information is buried in the data. This makes it necessary to analyze the data at *different scales*, a concern that is particularly relevant for larger data that span several magnitudes.

An interactive solution to this problem is multi-scale data exploration via zoomable interfaces [Bed11]. Zoomable visualizations enable the user to explore *different parts* of the data at *different scales*. Shneiderman formulated the *visual information seeking mantra* as a fundamental guiding principle for studying data at varying scales [Shn96]:

"Overview first, zoom and filter, then details-on-demand."

Shneiderman, 1996

The mantra suggests to start from an overview. The overview offers a big picture by visualizing as much data as possible. Yet it is only a coarse big picture, as there is no space for details. Detailed information can be gathered by zooming into subsets with less data. Less data means more space for details and closer inspection. With a zoomable visualization, the user can freely explore information at variable scales. The findings made during the exploration are compiled into a comprehensive understanding, very much like putting together the pieces of a puzzle.

Superficially, zooming sounds very much like successively focusing on data subsets D^+ as discussed in the previous section. The Shneiderman mantra, though, explicitly distinguishes between zoom and filter, since there is a big

REGULAR
VISUALIZATION
V(D)

FILTERING
V(D⁺)

ZOOMING
V⁼(D⁺)

Figure 4.14 Using $V^=$ to scale relevant data to fit the display space.

difference in how D^+ is represented visually: Zooming works with a visual encoding $V^=$ that is capable of scaling the graphical objects representing D^+, whereas with filtering, the focused subset is visualized only regularly $V(D^+)$. To illustrate the difference, Figure 4.14 compares a filtered view to a zoomed view. Filtering and zooming are similar in that no irrelevant data are displayed. Yet, for zooming $V^=(D+)$, the relevant data are scaled up so as to fit the available display space, the bars are enlarged and distributed evenly across the horizontal axis.

Obviously, the size of D^+ determines the scale of the visual representation. Discarding only a small part of the data such that $|D^+| \sim |D|$ will lead to only a marginal change in scale. Focusing on a very small portion of the data $|D^+| \ll |D|$ will bring the data analysis to a much finer scale. At any time, though, the user will see only a particular subset at a particular detailedness. It is the flexible interactive adjustment of focus (which part) and scale (which granularity) that enables users to engage in multi-scale data exploration.

4.5.1 Basics and Conceptual Considerations

Conceptually, zoomable visualizations build upon the notions of a *world*, a *viewport* defined in the world, and a *screen*. The world corresponds to the spatial arrangement of the visualized data. The viewport acts as a window into the world, and as such, determines what information is to be projected onto the screen. By moving the viewport, different parts of the world can be made visible. By resizing the viewport, one can adjust how much of the world is shown. The duo of move and resize operations of the viewport is commonly denoted as *pan and zoom*. Figure 4.15 illustrates a basic visualization world and how three different viewports lead to three different representations on the screen.

At the very heart of zoomable visualizations is the function $V^=$ that defines how the data representation gets adjusted when the scale is changed. This function can be implemented in different ways to accomplish:

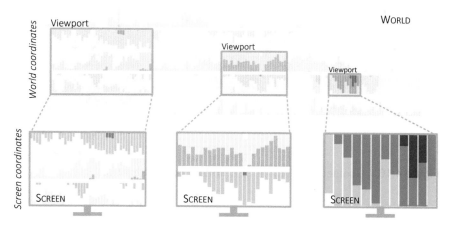

Figure 4.15 Illustration of the conceptual model of zoomable interfaces.

- geometric zooming or

- semantic zooming.

For *geometric zooming*, scale adjustments are considered at a geometric level only. That is, the visual representation is scaled according to the defined viewport. This can be done with a basic mathematical projection of the graphical objects. *Semantic zooming* goes beyond graphical scaling and allows for any kind of adjustments of the visualization depending on the scale. The additional semantics can be very valuable for visual data analysis.

A simple example from graph visualization can illustrate how dramatic the difference between a purely geometric zoom and a semantically enhanced zoom can be. Figure 4.16 contains a sequence of three zoom operations toward a central node of a graph (shown in dark green). The zoom is purely geometric: Everything becomes bigger as we zoom in. However, this does not help us much from a data analysis perspective.

Figure 4.17 shows a semantically enhanced zoom. The scaling takes effect only on the nodes' positions, but deliberately not on their size. The benefit of not scaling node sizes is that dense parts of the graph layout are untangled, creating an unobstructed view on the edges between the nodes.

There is a second reason for keeping node size constant. In our case, node size encodes the node degree. To ensure a consistent interpretation of the data, the node size should not be changed during zoom operations. Since many visualizations use size as a visual variable to encode the data, it is important to pay close attention to what should and should not be scaled when zooming.

The basic ingredients of zoomable visualizations are clear now. Next, we will look at the visual interface and the interaction that facilitate multi-scale data exploration.

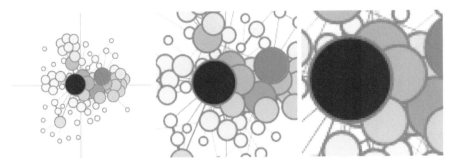

Figure 4.16   Geometric zooming of a node-link visualization.

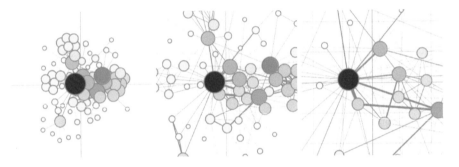

Figure 4.17   Semantically enhanced zooming of a node-link visualization.

4.5.2   Visual Interface and Interaction

At any time, a zoomable visualization shows only a particular view of the data. Therefore, zoomable visualizations have to be designed such that users can orient themselves and take navigation steps easily, allowing full concentration on the data analysis objectives. This requires a suitable visual interface as outlined below. The interface should support the user in coping with three questions in particular [Spe07]:

- Where am I?

- Where can I go?

- How do I get there?

Where Am I?   To facilitate the data exploration, it is essential to clearly communicate where the current view is located in the global context. Scroll bars as illustrated at the bottom and to the right of Figure 4.18 indicate where in x-direction and y-direction the current view is located in the world. From the size of the scrollbars, we can further infer how much of the visualization is covered by the current view. Scroll bars require relatively little display space, but it takes some mental effort to extract the information they bear.

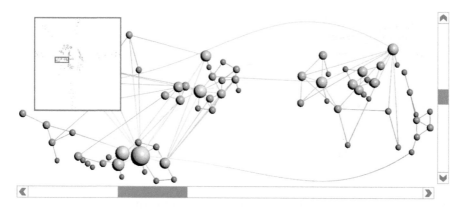

Figure 4.18   A zoomable graph visualization and its controls.

The overview+detail concept, as already described in Section 3.1.2, allocates more visual resources to convey information more explicitly. As can be seen in the left part of Figure 4.18, an overview shows a miniature version of the visualization in a dedicated window. Within the overview, the current viewport is marked with a red rectangle. This makes it quite easy to interpret the current view's location and scale in relation to the entire visualization. However, a disadvantage of embedded overviews is that they obscure parts of the actual visualization.

Where Can I Go?   Similar to how scrollbars and overviews communicate location and scale of the current view, they indicate where one could potentially go. The free space left and right as well as above and below the scrollbars and in the overview stand for parts of the data that can be accessed.

Yet, indicating where one could *potentially* go is only a first step. As a second step, it makes sense to hint at where one could *usefully* go. The more information about the world is provided, the easier it is for the user to decide on the destination to explore next. Therefore, overviews often embed a highly abstracted visual representation of the data. In our case, the overview depicts the global graph layout only via tiny colored dots. But even though there are neither edges nor any details in the overview, we can still see accumulations of nodes that may be worth visiting. The size of these accumulations suggests to us at which scale we should look at them.

How Do I Get There?   Knowing where to go, the next question is how to actually get there? Getting to a new view may be accomplished by *moving* and *scaling* the current view, or by *defining* a completely new one.

Conceptually, there are two basic options for users to actually carry out these operations. The first is to interact directly on the view. The second is to manipulate graphical objects of the interface. Table 4.3 lists some common

TABLE 4.3   Interactions for moving, scaling, and defining a zoomed view.

Direct interaction on the view	
Move view	Drag world in view, also known as *panning*.
Scale view	Mouse wheel, mouse drag, or pinch gesture on view.
Define view	Draw elastic rectangle in view.
Manipulation of interface objects	
Move view	Drag scrollbars or red frame in overview.
Scale view	Drag edges of scrollbars or red frame in overview.
Define view	Draw elastic rectangle in overview.

implementations for both options. Interacting directly on the view is good for making smaller changes to explore the immediate vicinity of the current view. However, substantial changes are more difficult to make because they would require many repetitive interactions. This is where the manipulation of interface objects can be more practical. Yet, it is important to realize that the precision with which can be interacted is limited. This is due to the fact that the interface objects represent the entire visualization space in a relatively small display space.

In summary we see that a suitable interactive visual interface is indispensable for multi-scale data exploration with zoomable visualizations. Yet, it does not stop there. Additional interaction aids and visual cues can further improve the utility of zooming and facilitate particular analysis tasks.

4.5.3   Interaction Aids and Visual Cues

With the techniques introduced so far, the navigation in the data is based solely on the visual layout of the data. For example, when we pan in a node-link diagram, we will get to see the data that is spatially near in terms of the graph layout. But what if the data analysis involves questions regarding the neighborhood in terms of the graph structure. The structural neighbors are not necessarily in the vicinity of the current view, but might be anywhere in the graph layout. The question is where are they and how can we get there?

Another common task is to look for data that are similar to those shown in the current view. Again, the similar data are not necessarily spatially near the current view, but could be in a completely different part of the visualization. Again we ask ourselves, where should we go and how can we get there?

Both scenarios outlined above have in common that the data we wish to inspect may not be visible on the screen. Therefore, our first goal is to make users aware of off-screen data. As a second goal, we want to enable users to actually navigate to these data. Third, we are interested in making the view change understandable. Next, we will see how off-screen visualization, navigation shortcuts, and animated feedback can help us achieve these goals.

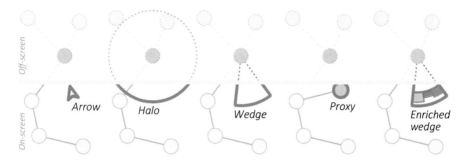

Figure 4.19 Visual cues for pointing to off-screen data. Adapted from [GST13].

Off-screen Visualization

Off-screen visualization is a form of focus+context representation where the zoomed view (the focus) is enhanced with additional context information. The idea is to embed additional visual cues into the display to make residue of otherwise invisible data elements visible.

Figure 4.19 illustrates examples of visual cues for off-screen nodes (dashed lines) in a node-link diagram. From left to right, there are five alternative visual cues. An arrow is a most simple mark to point at off-screen data. While arrows only indicate a direction, halos and wedges visualize direction plus distance [BR03; Gus+08]. By mentally completing the halo to a full circle, we can infer the full circle's center and so the location of an off-screen node. With a similar mental effort, we can determine a node at a wedge's tip.

Proxies aim at representing the actual off-screen data more faithfully [FD13]. Note how the proxy in Figure 4.19 allows us to see the off-screen node itself and also its connection to the visible part of the graph layout, rather than indicating direction and distance. Enriched wedges strive to balance several communicative goals [GST13]. They visualize direction, distance, and properties of the off-screen object. Even additional meta-information can be embedded, for example, to explain *why* a node is considered relevant.

Apparently, pointing to off-screen data has to be done sparingly to avoid cluttering the display. Therefore, an off-screen visualization is typically backed by a mechanism that automatically infers which parts of the data are potentially relevant. Such mechanisms can be based on degree-of-interest (DoI) functions or analytical calculations on the data. More information on DoI-based exploration of graphs can be found in Section 5.2.1.

Navigation Shortcuts

The next question to be addressed is how to get to data that are far from the current view. While basic pan and zoom interactions can of course be used for this purpose, we are now interested in making the navigation more cost-efficient. This can be achieved by means of navigation shortcuts.

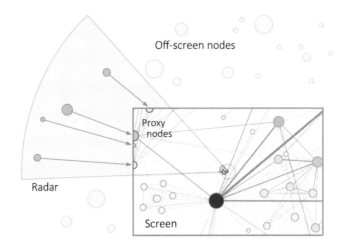

Figure 4.20 Bring & go with radar view and proxy nodes.

A navigation shortcut has a target in the data and is further associated with a viewport. The viewport is typically centered on the target, and its extent defines how much context around the target will be shown.

In order to actually take advantage of navigation shortcuts, we need to make them visible. A particularly useful approach is to employ the off-screen visualizations described before. The only thing that needs to be done is to turn the otherwise passive visual cues into active interface elements that can be interacted with. Navigating to an off-screen target is then as easy as clicking or tapping a visual cue.

The idea of combining off-screen visualization with navigation shortcuts is also known as *Bring & Go*. It is a very effective means to reduce navigation costs when exploring large data [Mos+09; TAS09]. Let us briefly illustrate the *bring* and the *go* part with an example.

Figure 4.20 shows a graph visualization with many nodes outside the current view. A so-called *radar* tool *brings* in the otherwise invisible graph nodes by placing proxy nodes at the screen boundaries. The proxies act as visual cues and also serve as navigation shortcuts. In order to *go* to a particular node, the user simply activates its associated proxy.

Using navigation shortcut like this is much easier than repeatedly performing pan and zoom operations. Note, however, that navigation shortcuts only work when potential targets are known. For exploring uncharted terrain, the basic pan and zoom operations are still important.

We are now familiar with several interactions that allow us to navigate from one portion of the data to another. Closing the action cycle, we will next consider animated visual feedback to make transitions between different views easily comprehensible.

Animated View Transitions

The most simple form of visual feedback is to instantly refresh the visualization once a new viewport is set. This corresponds to *static* visual feedback as introduced in Section 4.3.2. For small incremental changes of the viewport, as they occur when panning the view directly or when dragging a scrollbar, this type of visual feedback is a suitable option.

However, when the viewport has changed substantially, for example, after activating a navigation shortcut, simply replacing the old view with the new one may confuse the user. The problem is that the user has no chance of creating a mental connection between old and new view. This problem can be addressed by providing *animated* visual feedback. A smoothly animated transition makes it possible for users to comprehend how one view evolves into another, and hence, to stay oriented within the data.

Smooth viewport animations are based on the following general idea: First, *zoom out* from the current view, second, *pan* toward the new view's location, and third, *zoom in* to the reach the new view's scale [vWN04]. But instead of taking these steps one after the other, they are smoothly intertwined. It turns out that the math involved in a smooth viewport transition is not trivial. The interested reader is referred to the excellent mathematical derivation in the original paper [vWN04]. Here, we can only sketch the basic idea and illustrate it with an example.

For this purpose, let us take a look at a sequence of snapshots of a zoom animation. Figure 4.21 provides an overview with all intermediate viewports marked as gray rectangles. The animation starts with the smallest viewport in the bottom-left part of the visualization. The destination is a small subgraph at the top-right. In the first phase of the animation, we are taken to a zoomed-out

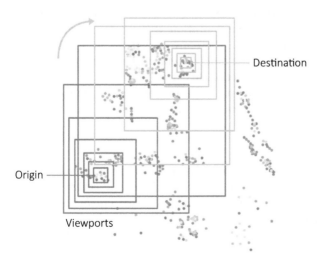

Figure 4.21 Viewports during an animated transition.

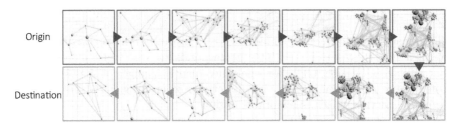

Figure 4.22   Snapshots of the viewport animation outlined in Figure 4.21.

view. Then, the view is smoothly panned and zoomed toward the destination. The result shown in Figure 4.22 is not only an eye-pleasing animation, but also helps users understand the view transition. Try for yourself to comprehend it without the overview and the intermediate steps of the animation. Looking at the origin and the destination alone, it is hardly possible to reach the same level of understanding as when following the animation.

We now have reached the point where the action cycle is closed. We have described several interactive mechanisms and corresponding visual means that allow us to explore data at varying scales. Common to all our previous considerations is that we worked with a two-dimensional zoom, which is a perfect match for the many two-dimensional visualization techniques in existence. In the following, though, we will see that one-dimensional or n-dimensional zooming is also important in data analysis scenarios.

4.5.4   Beyond Zooming in Two Dimensions

Multi-scale exploration via zooming as described above inherently relies on a two-dimensional visual representation of the data. What we actually explore is the 2D view space. Yet, there are situations in which it makes sense to make the exploration more independent of the view space, that is, to provide specific pan and zoom functionality depending on the data characteristics. In this section, we will look at the particular case of multi-scale exploration of univariate (1D) and multivariate (nD) data by the example of time series.

Exploring 1D Time Series

Understanding data in their temporal context is a particularly important analysis objective [Aig+11]. As we explore time series, we need to flexibly adjust *where* to look in time and also at *how much* we look. Range sliders are commonly applied to support one-dimensional navigation along the time axis. A range slider consists of a scale representing the time domain and two handles that define the time period to be visualized.

Figure 4.23 shows a range slider in combination with a spiral visualization. The slider's handles can be adjusted to narrow or widen the time period visible

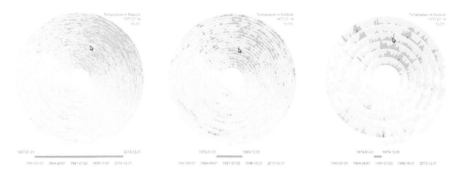

Figure 4.23 A range slider controls the time period mapped to a spiral visualization of the daily average temperature for the city of Rostock.

at the spiral. This corresponds to zooming in and out. A rather wide period facilitates an overview of the data (left spiral), whereas a rather narrow period allows us to see details (right spiral). Moving through time is possible by dragging the range marked in between the two slider handles. This corresponds to a pan operation. As the user manipulates the slider, different or more or less time steps are visualized, while the general spiral layout is maintained. It is left as an exercise to the reader to imagine what visual results a normal 2D zoom would produce.

In our example, zoom and pan are deliberately one-dimensional. By making the zoom independent from the 2D visual representation and hooking it up more tightly to the dimension of time, it is possible to define a time period of interest more directly.

However, there is a problem with sliders when used for navigating in time. Time-oriented data often contain thousands of time steps. The time series from Figure 4.23 consists of about 25,000 days worth of data. We would need a slider that has a width of 25,000 pixels to be able to guarantee that any date in the range can be accessed. If we had the 25,000 pixels, each pixel would represent exactly one date. Typically, however, we do not have so many pixels. More realistic are scenarios where our slider has a width of 1,000 pixels, meaning that 25,000 dates are mapped to only 1,000 pixels. So, by moving a slider handle by one pixel, we will not get to the next day, but to the next 25th day, essentially skipping 24 dates in between. There is no way we can access any of these 24 dates by direct manipulation of the slider. In short, when the data dimension we wish to navigate with a slider is reasonably large, some values could be inaccessible. This problem can be tackled by considering multiple scales not only for the visualization, but also for the interaction.

Multi-scale Input

The question is how to explore a larger range of values quickly and precisely by using a single continuous interaction gesture? With standard controls we

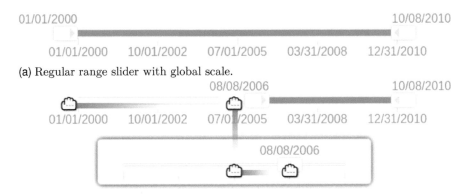

(a) Regular range slider with global scale.

(b) A slider with increased precision is dynamically added to the range slider.

Figure 4.24   Adjusting a time period at different input scales.

can either be fast (cover large distances) or precise (get to an exact spot), but not both. The reason is that the scale at which the interaction takes place is fixed. The time slider from Figure 4.23 facilitates quickly browsing through thousands of days of data. Yet, how can we navigate with precision, say to an exact date?

Interacting at dynamically changing scales is an answer to this question. Adding dynamic scaling to the input equation allows us to be swift and accurate at the same time. At a coarser scale, large distances can be covered, while at finer scales, interaction precision is increased. We will illustrate this with a simple example.

Let's assume our data domain covers the dates between January 1, 2000 and December 31, 2010. The goal is to focus on the subrange from August 8, 2006 to October 8, 2010. Figure 4.24a shows a regular range slider, where the upper limit has already been set as intended.

In order to set the lower limit, the user performs a continuous gesture as illustrated in Figure 4.24b. The gesture starts by grabbing the left slider handle, which is then dragged coarsely to the desired lower limit. As the exact date cannot be accessed directly, the user carries out a downward movement orthogonal to the slider. This triggers the dynamic appearance of an on-demand slider, where the gesture continues with a horizontal movement.

The important detail about the new slider is that it offers a higher precision. This is accomplished by mapping a smaller range of values to the slider. It covers only a local interval around the value where the cursor left the regular slider. Thanks to the increased precision, the new slider's handle can be moved exactly to the desired date. As this happens, the lower limit at the regular slider is updated automatically. Once the exact date has been reached, the gesture can be released, which dismisses the on-demand slider.

As we can see, a two-scale mechanism enables us to interact coarsely (with the regular slider) and precisely (with the on-demand slider) in one gesture. By incorporating additional increased-precision sliders, it would also be possible to interact at more than two scales. Such a general approach would be useful for exploring very large time series.

The example in this section started out with the goal to adjust a time period to specific limits. Some readers may argue that knowing the limits in advance, we could have set them simply by using the keyboard or a calendar widget. Yet, when exploring unknown data, it is often unclear where to look for interesting findings. They appear and disappear as we form and falsify hypotheses about the data that unfold before us. It can very well be that we spot something promising while moving the slider. This may change our course of action and we decide on the fly that studying a different subrange could be more rewarding. This is a major advantage of continuous interaction: With a single gesture, we can dynamically explore the data at flexible pace and precision. Such a fluid exploration is hardly possible by querying the data in a discrete fashion, be it via keyboard input or a calendar widget.

In the next section, we will continue to explore zoom concepts for data exploration. Yet, we will no longer navigate along the dimension of time alone, but rather expand zooming and panning to any dimension in the data.

Zooming Multiple Data Variables

Previously, we considered sliders as useful elements of the graphical interface. Next, we will see how a tight integration of sliders and visualization can support multi-scale exploration of nD time series.

Let us illustrate this for the example of axes-based visualizations. As you may recall from Section 3.2.3, the basic visual component of axes-based visualizations are axes, each being associated with a specific data variable and arranged according to a specific layout. The actual data representation is achieved by lines or dots placed between pairs of axes. With this general approach, a variety of visualizations can be generated, including parallel coordinates plots, scatter plot matrices, line charts, and TimeWheels.

An example of a TimeWheel is shown in Figure 4.25a. The central axis represents time, whereas the axes in the periphery represent time-dependent data attributes. Colored lines connect the time axis to all dependent attribute axes. Different colors are used to differentiate the attributes. The question we want to investigate is how can users flexibly explore not only time, but also the time dependent attributes via nD panning and zooming?

An answer is to make the axes of the TimeWheel interactive [TAS04]. As axes already represent ranges of observable values, we only need to incorporate facilities that allow us to adjust *what* and *how much* of the value ranges gets visualized. To this end, a slider is integrated with the axis as shown in Figure 4.26. The slider body and its two handles to the left and the right can be manipulated directly via drag gestures as usual.

Adjusting the slider has two complementary effects. First, the slider body marks the subrange of interest $D^+ = [D_{low}, D_{high}]$ within the global value

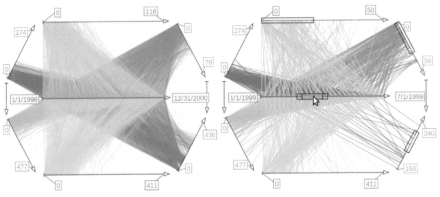

(a) Plain non-interactive axes.   (b) Axes with integrated sliders.

Figure 4.25   Integrated sliders for nD pan and zoom in the TimeWheel.

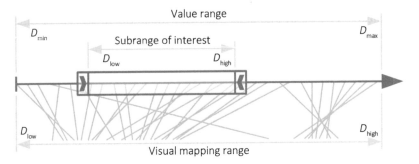

Figure 4.26   Integrated range slider for per-axis pan and zoom.

range $D = [D_{min}, D_{max}]$. Second, the actual visualization mapping of the axis
is altered. Instead of showing the global min-max range, only the subrange of
interest is visible. Note that the subrange extends over the entire length of the
axis. This way, we obtain a per-axis pan and zoom functionality. Narrowing
the slider allows us to look closer at details, whereas widening the slider will
take us back to a broader view. Moving the slider will bring us to a different
part of the value range.

An example with four zoomable axes is shown in Figure 4.25b. The time
axis has been zoomed to the first half of 1999, the two greenish axes have been
set so as to get rid of outliers, and the red axis focuses on a user-chosen value
range.

With these per-axis zoom interactions, we conclude our excursion into the
area of zoomable visualizations. As we have seen, there exists a rich set of
interactive tools and corresponding visual cues for comprehensive multi-scale
data exploration. In the next section, we will study interactive lenses as another

powerful concept for interactive data exploration. In contrast to pan and zoom, which typically affect the visualization globally, interactive lenses are tools for lightweight local adjustments of the visualizations.

4.6 INTERACTIVE LENSES

The interaction techniques discussed in the previous section enable us to explore different parts of the data, that is, to change *what* is shown on the screen. Another aspect of visual data exploration is to experiment with different visual encodings of the data, that is, to change *how* the data are visualized. In a sense, we broaden our view of interaction from exploration of the data space to exploration of the visualization space. This includes adjusting the mapping of data values to visual variables and the arrangement of visual marks on the display as explained in Chapter 3.

The standard way of supporting exploratory visualization adjustments is to provide a graphical interface with all kinds of control components. In Figure 4.3 back on page 140, we already saw such an interface. Altering parameters in the interface leads to a *global and permanent* change in the visualization. For example, when we switch the color scale, our visualization will be freshly painted overall.

An elegant alternative to the aforementioned standard approach are interactive lenses [Tom+17]. Lenses are lightweight exploration tools that can be added to a visualization on demand. Lenses can be used for various visualization adjustments, be it to encode data differently, to reconfigure the data's visual arrangement, to filter data according to certain conditions, or to connect related findings. A key characteristic is that lenses produce *local and transient* changes in the visualization. That is, the visual representation is adjusted only in selected parts and its original state is restored once the lens is dismissed. For example, a lens could be used to enhance the color coding to inspect details of local accumulation of data elements.

4.6.1 Conceptual Model

A schematic depiction of an interactive visualization lens is given in Figure 4.27. A lens is defined by its position, size, shape and orientation and divides the visual representation into an interior and an exterior part. Conceptually, lenses combine two interactions in a single tool: (i) interactive selection and (ii) adjustment of the visualization. A corresponding model can be defined based on the visualization pipeline. As you may remember from Chapter 2, the visualization pipeline describes how data are transformed to a visualization image via analytical and visual abstractions.

A visualization lens can be understood as an additional lens pipeline that is attached to a standard visualization pipeline as illustrated in Figure 4.28. The standard pipeline (bottom) produces a regular visualization. The lens pipeline (top) implements a *lens function* that generates a lens effect. There

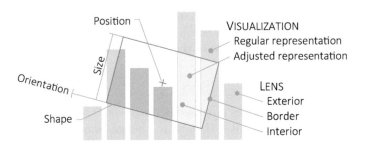

Figure 4.27   Schema of an interactive lens. Adapted from [Tom+17].

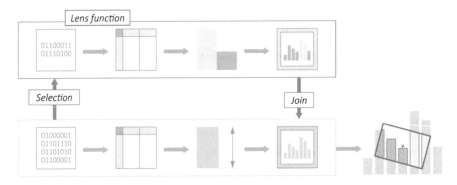

Figure 4.28   Model of a lens pipeline attached to a standard visualization. Adapted from [Tom+17].

are two points of information exchange between the standard pipeline and the lens pipeline. The first is a *selection*. It defines what is to be processed by the lens function. The second is a *join*, which specifies how the result of the lens function is to be integrated back into the standard pipeline. Next, we will describe these main ingredients of lenses in more detail.

The Selection   The selection corresponds to the content shown underneath the lens. Any type of content that is available along the visualization pipeline can be selected. A lens can directly select pixels from the image space. Content from other stages of the visualization pipeline can be selected as well, for example, a set of 2D or 3D graphical objects, a group of data elements, a range of values, or any combination thereof.

Typically, the selection will be a proper subset of the data that is significantly smaller than the original data. This allows a lens to perform calculations that would take too long for the entire dataset or would not be possible at all. As we will see a little later in this section, some lenses install mechanisms that automatically restrict the selection to maintain their operability.

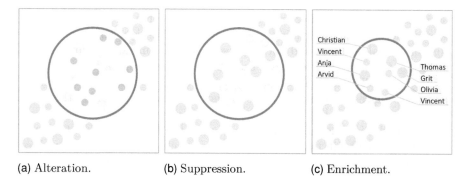

(a) Alteration. (b) Suppression. (c) Enrichment.

Figure 4.29   Fundamental effects of lens functions.

The Lens Function   While the selection specifies *what* is to be affected, the lens function defines *how* the visualization is modified. For example, when our goal is to re-color parts of the visualization, the lens function may achieve this effect by altering selected graphical objects. A selection that contains raw data opens up the possibility to create an entirely different visual representation, but restricted to the selected values.

The output generated by the lens function will lead to an alternative or modified visualization. In general, the goal is to improve the visualization with respect to the task at hand. To this end, a lens function can *alter* existing content, *suppress* irrelevant content, or *enrich* with new content. Figure 4.29 illustrates the different options. In Figure 4.29a, the visual encoding inside the lens is altered to emphasize the small dots. In contrast, the lens in Figure 4.29b suppresses the small dots deemed less relevant. Finally, Figure 4.29c shows a lens that enriches the visualization with labels for dots within the lens interior [BRL09].

The lens function usually depends on parameters that control the lens effect. A magnifying lens, for example, may expose the magnification factor as a parameter. A filtering lens may be parameterized by a threshold to control the amount of data to be filtered out. In general, a lens may be parameterized by an alpha value used for blending lens and visualization when both are joined.

The Join   To create the final image, the lens effect needs to be joined with the base visualization. Traditionally, the visible effect of a lens is confined to the lens interior. In the context of visualization, though, it can be practical to allow lenses to affect the visualization beyond their interior or even show their effect separately.

Yet, most lenses follow the metaphor of conventional lenses where the visual effect manifests exclusively in the lens interior. This can be accomplished with the following generic three-step procedure. First, render the base visualization, optionally sparing the interior of the lens. Second, fetch the lens effect to the lens interior, optionally blending with the base visualization. Third, incorporate suitable visual feedback and optional user control elements. In our examples, a thick outline makes clear to the user that a lens is in operation.

Conceptually, the join can be done at any stage of the pipeline. If the join takes place at the early stages of the visualization pipeline, the visual effect may go beyond the lens. For example, a lens may adjust the positions of selected nodes in a node-link diagram. As a side effect, the incident edges of the altered nodes will take different routes as well, which in turn introduces (limited) visual change into the base visualization outside the lens.

In summary, selection, lens function, and join describe the key components of lenses. Conceptually modeling lenses as secondary visualization pipelines makes it not only possible to use multiple lenses in the same visualization, but also to combine different types of lenses to create composite lens effects. Later in Section 4.6.3 we will see an example. Next, our attention shall first be drawn to the properties of lenses and means of adjusting them.

4.6.2 Adjustable Properties

From the perspective of a user who is actually working with a lens, two questions are relevant: *What* properties of lenses exist and *how* can they be adjusted to suit the user's data analysis objectives.

Lens Properties

Looking at lenses, their geometric properties are the first to catch the eye. Position and size of a lens are most relevant. They determine where and to which extent a lens takes effect. Another prominent property is the lens shape. Following the classic prototype of real lenses, many virtual lenses are of circular shape as shown in Figure 4.30a. Rectangular lenses are common as well. For non-circular lenses, also the orientation is relevant. Orientable lenses can be better adapted to the underlying data as illustrated in Figure 4.30b.

In addition to the geometric properties, there are the parameters that control the inner workings of lenses. We already mentioned magnification factors and filter thresholds as examples of lens parameters. In general, lens parameters are often used to balance the *strength* of the lens effect, where strength can have different meanings, for example, how much more detail is added, how much irrelevant data are suppressed, or how substantially the base visualization gets altered.

Interactive and Automatic Adjustment of Lenses

A great deal of the flexibility attributed to lenses pertains to the possibility to adjust them interactively via direct manipulation. Complementary automatic mechanisms may support the adjustment of lenses.

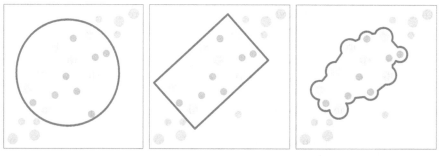

(a) Circular.　　　　　(b) Rectangular orientable.　(c) Content-adaptive shape.

Figure 4.30　Lenses with different shapes and orientation.

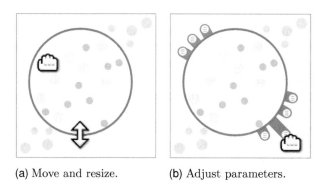

(a) Move and resize.　　　　(b) Adjust parameters.

Figure 4.31　Direct manipulation of lenses.

Direct manipulation as shown in Figure 4.31a is the preferred way of interactively adjusting position and size. Both properties can also be set automatically. For example, one can position a lens automatically at interesting data tuples. Automatic adjustments of the lens size can help to cope with the computational costs for producing the lens effect and also the cognitive costs for making sense of it. An example is the labeling lens from our earlier Figure 4.29c. When being moved into dense parts of the data, the lens automatically reduces its size to limit the number of labels. This way, the lens can keep the labels readable and the algorithmic runtime low [BRL09].

Interactively switching between different lens shapes is possible, but rather uncommon. More interesting are lenses that adapt their shape automatically based on characteristics of the data [Pin+12]. Such self-adapting lenses are particularly useful in cases where the lens effect needs to be confined to complicated geometric features in the visualization. As an example, Figure 4.30c illustrates a lens that has adjusted itself to a cluster of data elements.

Finally, we need to think about the internal parameters of lenses. For occasional adjustments, it is sufficient to rely on standard widgets. Parameters that need to be fine-tuned more frequently are preferably adjustable via dedicated mechanisms. Figure 4.31b shows an example where custom interfaces elements are provided directly at the lens [KRD14]. There are also lenses that can adjust their parameters automatically to the data. An example is to tune a sampling rate parameter to the data density underneath the lens [ED06].

As we see, the adjustable properties of lenses (geometry plus parameters) make them very flexible data exploration tools. Lenses can be easily controlled via direct manipulation or through automatic procedures that adapt the lens to the underlying data. How lenses can be applied to actually accomplish visual analysis tasks will be demonstrated next.

4.6.3 Lenses in Action

So far we have considered lenses on a rather abstract conceptual level. In this section, we will illustrate the versatility and utility of interactive lenses in the context of visual data analysis scenarios. We will consider four practical problems and corresponding lenses to solve them. We will start with the quite common task of studying specific details in the visualization. Second, we will explain how lenses can support the exploration of structural relationships in graphs. Adding space and time, the third lens will help us understand temporal aspects of geo-spatial movement trajectories. Finally, we will make a step from altering the visualization to altering the actual data with an edit lens.

Exploring Details with a Fish-eye Lens

Magnifying glasses have an ancient history as tools allowing us to look at details that cannot be seen with the human eye alone. On a computer screen, interactive magnifying lenses serve the same purpose: They are positioned on the screen where a more detailed view is needed. The lens will then transform the content underneath it according to a mathematical specification.

A prominent example of such a mathematical specification is the *fish-eye distortion* [SB94]. It gradually pushes content from the lens center outward. As shown in Figure 4.32, this effectively magnifies the content near the mouse cursor and allows us to see the details there. Because the fish-eye distortion smoothly embeds the details within the global context, it is conceptually a form of focus+context representation as introduced in Section 3.1.2.

The fish-eye lens already demonstrates the utility of lenses for altering a visualization in a dynamic and lightweight manner. Next, we will elaborate on lenses that are particularly useful for exploring structures in graph data.

Exploring Structural Relationships with Graph Lenses

When exploring graph data, structural relationships play an important role. Node-link diagrams, as introduced in Section 3.5 allow us to see how nodes are

(a) Regular map visualization. (b) Details magnified with a fish-eye lens.

Figure 4.32   Magnifying details in a map visualization with a fish-eye lens.

connected and if there are any communities or clusters. Next, we will apply graph lenses to enhance exploratory work with node-link diagrams [Tom+06].

Let us take a look at the zoomed-in graph visualization in Figure 4.33a. We are interested in the edges that connect the node in the center. As we can see, edge clutter is a problem in our example. There are many edges and we cannot really say which edges do actually hook up to our node of interest and which are just passing by. An interactive lens can help us out. In Figure 4.33b, we use a *local-edge lens* to clear the visualization of irrelevant edges. The lens suppresses edges that do *not* connect to nodes inside the lens. We can now easily see that our node of interest has seven incident edges.

What we cannot see, though, are the adjacent neighbor nodes, which are beyond the current view. We could pan and zoom to each one of them, but this could take a while and we cannot really be sure what we will find. Using a lens can be more efficient in this situation.

This time, we apply a *bring-neighbors lens*. As its name suggests, the lens will bring to us the neighbors of the nodes inside the lens. As the lens is moved toward the node of interest, its neighbors will be gradually drawn toward the lens. When the lens is exactly on top of the node, all its neighbors will be inside the lens as we can see in Figure 4.33c. The lens effectively produces a local overview of the neighborhood of the nodes covered by the lens. There is no need to manually visit the neighbors, the lens brings them in for us to inspect them.

The bring-neighbors lens works well when the neighbors are evenly distributed across the graph layout. However, if this is not the case, we could end up with the majority of the nodes occluding each other at the lens center. To solve this problem, we can exploit the fact that lens effects can be combined to create composite lenses.

In our case, we combine the bring-neighbors effect with a fish-eye distortion. The combined effect will bring in the neighbors, but those that would accumulate too tightly at the lens center will be pushed outward to loosen

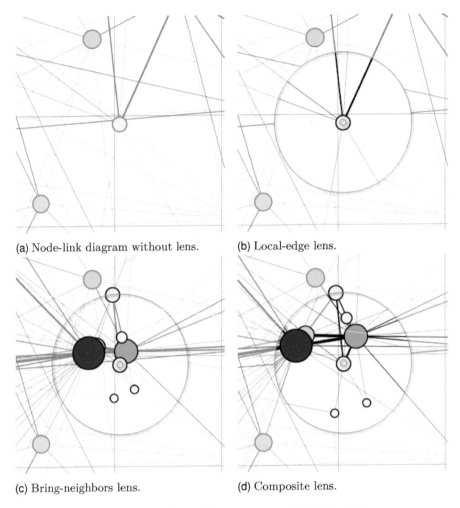

(a) Node-link diagram without lens.

(b) Local-edge lens.

(c) Bring-neighbors lens.

(d) Composite lens.

Figure 4.33   Graph lenses for exploring structural relationships.

the clutter. Figure 4.33d shows the result. Looking closely, you will realize that the figure actually shows a composite lens that combines all three effects mentioned before: local-edge effect plus bring-neighbors effect plus fish-eye effect.

We have just seen lenses in action for the specific task of exploring graph data. The lenses helped us tidy up edge clutter and peek at neighborhoods that are otherwise not visible at a glance. Next we will present a lens particularly designed for exploring temporal dependencies of movement data.

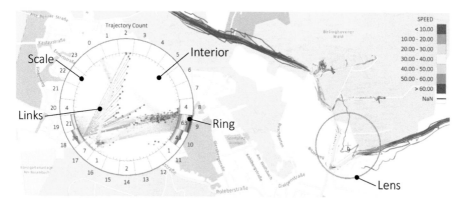

Figure 4.34 A lens to query temporal characteristics of movement data.

Exploring Temporal Aspects of Movement Data with a Lens

Back in Section 3.4.3, we discussed the challenge of visualizing spatio-temporal data. Showing space, time, and data attributes simultaneously at full detail is typically difficult. A more practical solution is to depict two aspects in full detail and add the third aspect interactively on demand for a selected part of the data. Here, we will enhance a visualization that focuses on space and attributes with an interactive lens to integrate the temporal aspect on the fly.

Our point of departure is a 2D visualization of spatio-temporal movement data as shown in Figure 4.34. A map provides the spatial context. Trajectories of moving cars are visualized as lines, where colors encode speed. What we can see from this kind of representation is *where* cars drive at certain speeds. The *time lens* will help us to also see *when* [Tom+12].

The lens effect is shown in an auxiliary circular display to the left in Figure 4.34. The *interior* shows a scaled copy of the trajectory points selected with the *lens*. The time-dependent speed attribute is visualized in the histogram *ring* around the interior. Our particular example reveals that movements in the selected region occur mostly around 9–10 and 18–19 o'clock, with the speeds being evenly distributed. Moreover, *links* connect the points in the interior with a finer-grained minute *scale*. This allows us to see additional details. For example, the trajectory marked in gray represents a movement that took place around 18:15 o'clock as indicated by the gray links accumulating at minute 15 of the 18th hour.

The insights gained with the time lens could not be obtained with the plain trajectory visualization alone. This demonstrates quite nicely how useful lenses can be when it comes to accessing additional information and details of complex data on demand.

All described lenses, the time lens, the graph lenses, and the fish-eye lens help users explore the data. Next, we go one step further and study a lens that supports editing the data.

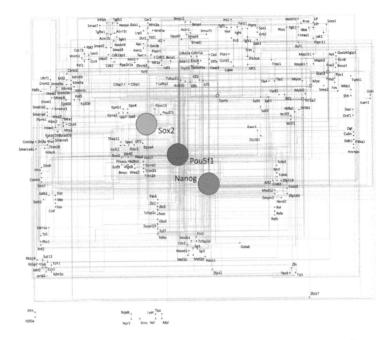

Figure 4.35 Orthogonal node-link diagram of a biological network.

Employing a Lens for Semi-automatic Graph Editing

During the exploration of data, one may stumble upon findings that make it necessary to correct the data, for example, to insert missing elements, update erroneous data values, or delete obvious outliers. The goal in this section is to demonstrate how a lens can support such an on-the-fly data editing.

Our particular example is about editing graphs. Suppose you have to insert a node with say a dozen of edges into a network such as the one shown in Figure 4.35. It is obvious that positioning the node and routing each of the edges by hand would be tedious work and take a lot of time. A graph layout algorithm could do the math and compute a high quality layout. But most algorithms would recompute the layout globally, which would harm the mental map that we or others might already have about the data. What we need is a tool that lets us edit the data locally without intensive manual labor and no global changes of the layout.

Again, we employ a lens to address this problem. Fitting to its purpose, the lens is called *edit lens* [Gla+14]. It supports three fundamental edit operations: insert, update, and delete. Figure 4.36 illustrates the lens being used for inserting, updating, and deleting a graph node. The only manual work required is to place the lens within the layout to specify where an edit operation is to take effect. In a sense, the adjustable lens acts as a coarse human-specified solution to be refined with automatic methods.

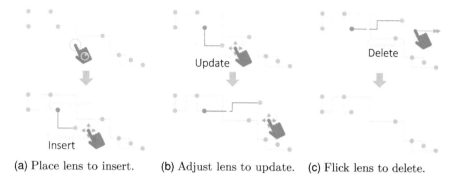

(a) Place lens to insert. (b) Adjust lens to update. (c) Flick lens to delete.

Figure 4.36 Editing using the edit lens. Adapted from [Gla+14].

The automatic part consists of two steps. The first step determines a suitable unoccupied area in the lens interior where the edited item can be placed. Second, the precise spot within that unoccupied area is computed based on heuristics for different graph aesthetics criteria, such as maximum distance to other nodes, short overall edge length, or low number of edge bends.

During the editing, the user is free to move and resize the lens and to choose a different heuristic. The lens will compute and suggest suitable node positions and edge routes on the fly. Only if the user agrees with a suggested solution is the result of the edit operation committed to the data.

To summarize, the edit lens simplifies fully manual editing to semi-automatic editing. This eases editing operations considerably, because the user only needs to define a coarse region interactively, rather than precise positions or routes. The algorithmic part of the lens computes suggestions for precise solutions, which the user can customize on-the-fly. Finally, the lens is integrated into the regular visualization so that data analysis and data editing can go hand in hand.

The edit lens concludes our journey into the world of interactive lenses in the context of visualization. In addition to the conceptual ideas behind lenses, we have described several exemplars of lenses for a number of different tasks, including looking at details, exploring graphs, incorporating temporal information, and even data editing. For more information on the described lenses, the interested reader is referred to the list of references collected at the end of this chapter.

This section discussed lenses as a versatile interactive approach to support data exploration and analysis. The focus was on one tool for many different tasks. In the next section, we will flip the perspective. The focus will be on one task for which a comprehensive set of interaction techniques will be introduced. The particular task we will be dealing with is visual comparison.

4.7   INTERACTIVE VISUAL COMPARISON

Visual comparison takes a central role during data analysis activities. By inter-actively comparing different parts of the data, users may formulate, confirm, fine-tune, or reject initial hypotheses, draw corresponding conclusions, and thus can gain a better understanding of the data.

Elementary comparison is often a predecessor to a more in-depth data analysis. For example, we can compare successive data values in a time series of stock prices to identify trends. Comparing the trends enables us to find groups with similar trend behaviors. Still more insight can be gained by comparing the groups, for example, to study if certain behaviors occur in specific periods of the fiscal year. It is also common to derive quantitative statements about the compared data to capture their degree of relatedness. A corresponding notion is that of similarity (or dissimilarity), which plays an important role in many higher-level knowledge generation activities.

This section deals with dedicated interaction techniques specifically designed to support visual comparison tasks. But before we look at these techniques in detail, we need to understand what comparison is and how visual comparison takes place.

4.7.1   Basics and Requirements

What precisely do we mean by comparison? Given individual data values p and q (or sets of values P and Q), *comparison* tasks are defined as the search for a relation $\mathfrak{r}$ such that $p \, \mathfrak{r} \, q$ (or $P \, \mathfrak{r} \, Q$). When comparing numerical values, order relations $\mathfrak{r} \in \{<, \leq, =, \geq, >\}$ are of great practical relevance. Specific relations exist for comparing temporal data (e.g., before, during, after) and spatial data (e.g., inside, overlap, touch).

There are three fundamental visual designs specifically for comparison tasks: *juxtaposition*, *superposition*, and *explicit encoding* [Gle+11]. Figure 4.37 provides a comparison (pun intended). Juxtaposition shows the data side-by-side, that is, in separate spaces. In contrast, superposition stacks data in a unified visual space. Explicit encoding means calculating and visualizing numeric differences of the data being compared.

However, comparing data visually without dedicated interaction support is typically non-trivial. For example, suppose you have spotted two interesting

JUXTAPOSITION   SUPERPOSITION   EXPLICIT ENCODING

Figure 4.37   Visual designs for comparison tasks.

patterns in different parts of a color-coded spreadsheet visualization. For comparing them, you first have to visit one pattern and memorize it. Then you have to navigate to the second pattern and compare that to the stored mental image of the first one. This procedure is inefficient, because it requires you to scroll over and over again, and is also error-prone, because the actual comparison is carried out based on a mental image from your short-term memory.

The previous statements suggest that visual comparison involves multiple actions working in concert to reach a higher level of analytic thinking. From an interaction perspective, visual comparison is a procedure that comprises three phases:

1. **Select** the information to be compared.

2. **Arrange** the selected information for comparison.

3. **Compare** the arranged information visually.

From these phases, we can infer important requirements. First, a set of comparison candidates has to be selected and maintained interactively. The number of candidates is usually small because our visual working memory is limited [PW06]. Yet, data may enter or exit the set of candidates on the fly as user interests change during the data investigation.

Second, the data to be compared have to be rearranged dynamically to facilitate their comparison. This is necessary because many standard visualizations are oblivious to comparison tasks and arrange the data according to some fixed layout algorithm or some naturally given mapping such as geographic positions. As a consequence, there might be larger gaps between the data to be compared, which make comparisons more difficult. The eyes have to look back and forth between different parts of the display frequently. Moreover, when studying larger data with zoomable visualization interfaces as introduced in Section 4.5, it is not guaranteed that all relevant data are visible at all. Many manual navigation steps might be necessary in order to successfully accomplish a comparison task. At the same time, the short-term memory has to store not only the locations of the data, but also their visual representations.

Finally, the actual comparison is performed. Juxtaposition, superposition, and explicit encoding form the visual basis for the comparison. However, it is not clear upfront which is the best strategy for the data at hand. Therefore, the user should be able to interactively choose and parameterize comparison strategies as needed.

In summary, we see that visual comparison is a highly dynamic procedure. Interaction techniques are needed to flexibly link the phases of comparison and to account for the changing interests during comparative analytic activities. In the following, we introduce dedicated interaction designs for visual comparison. Next in Section 4.7.2, our main priority will be naturalness of visual comparison. Later in Section 4.7.3, we will additionally consider aspects of cost efficiency.

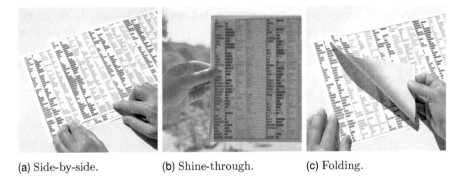

(a) Side-by-side.   (b) Shine-through.   (c) Folding.

Figure 4.38   Natural behavior of people comparing information on paper.

4.7.2   Naturally Inspired Comparison

Natural interaction is a theme of interaction research that focuses on enhancing interactive experiences by drawing inspiration from natural human behavior. Following this thinking, we first observe how people compare information naturally and then develop techniques that mimic people's natural behavior [TFJ12]. Figure 4.38 illustrates the three basic strategies that can be observed when humans compare information printed on paper:

(a) Side-by-side comparison: Sheets of paper are moved on a table until they are arranged side-by-side to facilitate comparison.

(b) Shine-through comparison: Sheets of paper are stacked and held against the light to let information shine through and blend.

(c) Folding comparison: Sheets of paper are stacked and individual sheets are folded back and forth to look at them in quick succession.

Our goal is to replicate these natural strategies. To this end, we have to design corresponding virtual counterparts for the involved visual components and interactive procedures.

In terms of the visual components, we need a virtual comparison workspace and a virtual equivalent for sheets of paper. The workspace will be a zoomable visualization space based on the ideas discussed in Section 4.5. Sheets of paper are replicated as visualization *views* that reside in the zoomable space. Within the zoom space, views can be moved freely in analogy to arranging pieces of paper on a table. How the visual components are put to use for naturally inspired comparison will be explained in the following.

Selecting What to Compare

The comparison procedure starts with the selection of data to be compared. When users spot something interesting for comparison, they can simply mark

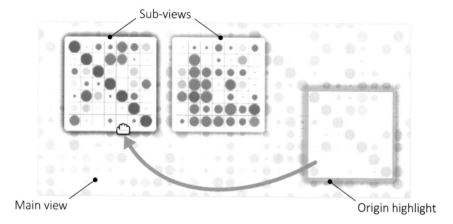

Sub-views

Main view

Origin highlight

Figure 4.39 Creating sub-views for comparison. A red frame indicates where the left sub-view has been detached from the main view.

it with an elastic rectangle. The system then creates a new view corresponding to the marked region, a sub-view of the entire visual representation so to say. Once created, sub-views exist as independent views in the visualization space.

The real-world analog would be to create a copy and cut it out to contain only the information of interest. Yet, in the real world, we would lose the connection between the cutout and its origin. On the computer, the parent-child relationship of views can be preserved in a view hierarchy. As a result, users no longer need to mentally keep track of what they want to compare because this information is now externalized in the form of dynamically created views collected in a view hierarchy.

The view hierarchy further makes it possible to embed visual cues for highlighting a sub-view's origin on demand. Figure 4.39 illustrates this for a matrix visualization from which two sub-matrices have been created. As the user points at the left one, a red frame indicates where the sub-matrix has been extracted from its parent matrix in the background.

Arranging for Comparison

The second comparison phase consists of arranging the views to be compared. Using simple drag gestures, views can be arranged in the visualization space similar to shifting paper on a table. In the real world, people use edges of papers or patterns on the table surface to guide the arrangement. In the virtual world, *snapping* can be used to assist in arranging views. Snapping automatically aligns views with respect to certain features in the visualization. This way, costly pixel-precise adjustments can for the most part be avoided. In our example from Figure 4.39, snapping helps to maintain the alignment of matrix cells.

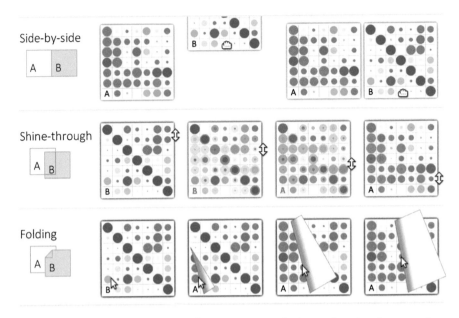

Figure 4.40 Overview of natural interaction techniques for visual comparison. Adapted from [TFJ12].

Comparing Views

Phase three is the actual comparison. Here, we take a simplified perspective on the underlying problem and abstract from the specific details of the data being compared. Chapter 3 made clear that plenty of visualization techniques are available to generate expressive visual representations of data. Under this assumption, we can simply resort to our views and sub-views as the objects to be compared visually. Next, we introduce the virtual analogs for the natural comparison strategies: side-by-side, shine-through, and folding. They are also summarized in Figure 4.40.

Side-by-side Comparison   Side-by-side arrangements provide us with a complete sight of the data, which is helpful as an overview. On the other hand, when comparing details, resolving spatial references among the views requires some cognitive effort. In particular, we have to move our eyes to check if a feature in one view can be found in the very same spot in another view.

For example, when we compare matrix cells at position $[i, j]$, we can follow along the i-th row easily from the left sub-matrix to the right. However, as we do so, we have to take care to stop correctly at the j-th column, which may be doable for our small example, but would certainly require some counting for a larger matrix. For more complex visualizations, the eyes might need to move to and fro multiple times to be sure of looking at the same reference spot.

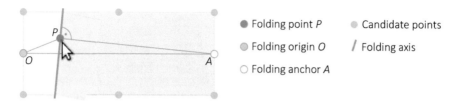

Figure 4.41 Folding geometry. Adapted from [TFJ12].

Shine-through Comparison As with natural comparison of paper, an alternative is to let views overlap. In reality, people often purposely stack papers on top of each other in order to create a unified comparison space. While spatial references are aligned then, mechanisms are needed to look *through* or *behind* the paper sheets. One way to resolve the occlusion is to hold the paper against the light. The degree to which information shines through can be controlled by altering the viewing angle with respect to the light source.

On the computer, shine-through comparison can be realized via alpha-blending. That is, views are made partially translucent, where the level of transparency can be varied by the user. Shine-through makes it easy to compare shapes and sizes in the graphical depiction of the data. On the other hand, blending the views implies that we mix colors as well, which is hindering the comparison of color-coded visual representations.

Using shine-through, it can also be difficult to figure out which view contributes its data to a particular feature in the blended image. Or put differently, shine-through favors a merged view on the data at the cost of losing separability of individual data elements. The folding interaction described next addresses this aspect.

Folding Comparison Folding paper back and forth allows us to compare information shown on different sheets. The same is possible with our visualization views: To uncover occluded information, the user can fold away or peel off views as if they were virtual paper. The folding resolves the occlusion temporarily, while otherwise keeping the views in place to preserve their arrangement.

In a data analysis setting, the fold should appear directly where the user's focus is, which is typically the location of the pointer cursor P. Knowing P, a heuristic can be applied to determine from a set of candidate points a folding origin O and a folding anchor A. For real paper, the folding origin O corresponds to the spot where we would grab a page for folding it. The anchor A represents the fixture around which the paper is folded, such as a staple or the binding. Finally, the folding axis is constructed as a line originating at P and being perpendicular to the line PA as illustrated in Figure 4.41.

The folded visualization view can be rendered in different styles. As shown in Figure 4.42, the styles are to vary in naturalness, information-richness, and degree of occlusion. For example, the visual effect resembling *natural* paper

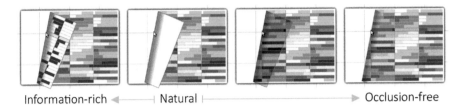

Information-rich ◄────┤ Natural ├────────────────────► Occlusion-free

Figure 4.42 Information-rich, natural, and occlusion-free folding styles. Adapted from [TFJ12].

folding leaves the folded backside blank. A more information-rich style would enhance the backside with additional information. An occlusion-free style could be restricted to only showing a subtle shading gradient.

For a natural feeling and realistic appearance, the folding is animated using a simple physically based spring-mass system. The animation starts with a spring force smoothly pulling the fold from the origin O to the pointer position P. When the user adjusts the pointer, P is updated so that the fold smoothly follows up. When the user releases the fold, the spring force is inverted to pull the fold back from the pointer P to the folding origin O, effectively unfolding the view. With the animated visual feedback, the virtual folding interaction for comparison tasks is complete.

It is time for a recap. We studied a comprehensive repertoire of techniques for interactive visual comparison. By dynamically extracting sub-views, it is possible to flexibly define what should be compared. Side-by-side, shine-through, and folding interaction enable users to carry out the actual comparison in different ways. The natural inspiration behind the interaction design contributes to making the comparison intuitive and easy. However, a pressing question remains: Under which circumstances can we apply which technique most effectively? Finding answers to this question requires further extensive research and elaborate user studies.

So far, we left an important advantage of interactive systems (compared to paper) out of consideration: the ability to do automatic computations, for example, to calculate differences and show them directly. In the next section, we will continue with interaction specifically for visual comparison. However, our focus will shift from naturalness to reducing the interaction costs for comparing data, also by including automatic calculations.

4.7.3 Reducing Comparison Costs

Before taking measures to reduce costs, we need to know where they accrue. In Section 4.2.1, we learned that interaction costs in general can be attributed to executing the interaction and to evaluating the visual representation. For comparison tasks, the following specific costs can be identified:

- Costs for selecting the data to be compared

- Costs for carrying out the comparison

- Costs for understanding the data in context

To reduce these specific comparison costs, we will now combine several of the techniques presented in the previous sections, including automatic selection mechanisms from Section 4.4, off-screen visualization and navigation shortcuts from Section 4.5, and the idea of local and lightweight adjustments as we know them from the lenses from Section 4.6. Let's start with reducing selection costs.

Semi-automatic Selection of the Data to be Compared

For printed non-interactive data visualizations, much of the costs for comparison are related to memorizing where relevant data are located and what characteristics they exhibit. On a computer, interactive selection enables the user to mark and highlight interesting data (see Section 4.4), which effectively off-loads otherwise mental effort to the machine.

As user interests change frequently during data exploration, it makes sense to invest in reducing the selection costs by integrating automatic mechanisms. The idea is to reduce the manual selection of n data elements (or data subsets) to be compared to only a single manual selection plus an automatic selection of the remaining $n - 1$ elements (or subsets).

The utility of this idea largely depends on defining a semantically meaningful automatic selection. A sensible approach in the context of comparison tasks is to rank the data according to their similarity (or dissimilarity). In other words, when the user selects a first data element, the $n - 1$ most similar (or dissimilar) elements are added to the selection automatically. There are a number of similarity measures for different types of data that can be used for this purpose. Euclidean or Manhattan distance work well for numerical data, Edit or Hamming distance operate on string data. Categorical data require dedicated distance measures [BCK08]. In cases where the similarity of complex data subsets must be captured, it makes sense to extend to multi-dimensional or subspace measures [Tat+12].

As an alternative to similarity-based automatic selection, one could consider mechanisms that traverse the internal (graph) structure of the data or utilize flexible degree-of-interest (DoI) functions, as for example explained in Section 5.2.1 of Chapter 5.

Irrespective of the method being employed to drive the automatic selection, the benefit for the user is significant: A single click (or tap) is enough to create a selection of n comparison candidates. This corresponds to a reduction of the costs from $O(n)$ down to $O(1)$. Of course, additional manual steps may be taken to complement or refine the automatic selection.

Dynamic Rearrangement and Visual Cues for Comparison

Once data have been selected for comparison, the next step is to compare them in detail. However, typically it will take a number of navigational steps until we have acquired enough information to draw a comparative conclusion. The goal is to reduce the time-consuming navigation between the data to be compared. Instead of us collecting the required information, the system should automatically *bring* it to where we need it. To this end, the data have to be rearranged dynamically.

The basic idea is to create a juxtaposition arrangement to facilitate the comparison on the fly. One option to do this is to form a ring, a so-called *CompaRing* [Tom16]. As shown in Figure 4.43, the CompaRing is a circular arrangement of *slots* to be filled with the data for comparison, regions of a choropleth map in our case. When the CompaRing is activated, the previously selected data are dynamically relocated from their original position to the slots. With all relevant data being now displayed at the ring, the comparison can be carried out more directly and more easily.

However, the relocated data are now detached from their original spatial context, which could be detrimental to other data analysis objectives. Therefore, *indicator arcs* point in the direction where a slot's data are originally located. Wide arcs (max. 90°) stand for far-away origins, whereas narrow arcs (min. 10°) suggest the origin is close.

The indicator arcs can further be exploited to augment the comparison with additional details. In order to make even subtle variations visible, the pairwise differences between a selected slot and all other slots of the CompaRing are calculated and color-coded into the indicator arcs. In Figure 4.43, a diverging red-blue color scale indicates negative and positive deviations from the slot under the mouse pointer.

The key benefit of the CompaRing is that it is no longer necessary to collect and memorize data characteristics, as the CompaRing brings the required information to us. Moreover, the explicit visual encoding of calculated local differences grants us insight into details that are not evident in the base visualization. This naturally reduces the costs for comparison tasks.

Understanding Data in Context with Navigation Shortcuts

So far, the CompaRing facilitates the plain comparison of the selected data. For more profound insight, we also need to understand the data in their spatial context. This still requires us to navigate to individual places in the data manually. In order to reduce the cost for navigation, the CompaRing picks up the idea of navigation shortcuts as introduced in Section 4.5.3. Each slot of the CompaRing serves as a trigger for a smooth animation that takes the user (and the CompaRing) to the corresponding data's original position. Manual navigation steps are thus reduced to a minimum.

On top of that, using the navigation shortcuts in combination with similarity-based automatic selection as described earlier enables a whole new

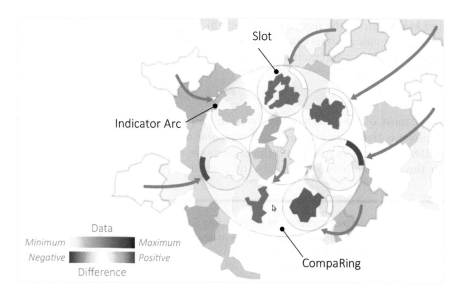

Figure 4.43 Relocating selected regions to form a ring for easier comparison. The map background has been desaturated for the purpose of illustration.

type of data-driven *bring & go* navigation. We can use a navigation shortcut to *go* to the context of some data of interest. If we find some interesting data in the context, we can mark them and the automatic selection will *bring* related data to the CompaRing and hence to our attention. Each of the newly brought data elements can then be used to *go* to yet another destination to continue the data exploration. The interesting thing about this type of navigation is the combined use of semantic (data similarity) and spatial (data context) relationships as paths through the data.

This concludes the section on interactive visual comparison. We learned that visual comparison is a high-level activity that involves several intermediate steps. In the first part of this section, we replicated human comparison behavior on the computer to facilitate natural comparative data analysis. Secondly, we presented several strategies to reduce comparison costs by complementing interactive methods with automatic mechanisms. As a result, we obtain a comprehensive repertoire of dedicated techniques for visual comparison. A take-home message of the described techniques is that visual comparison requires a high degree of flexibility when it comes to selecting data, rearranging their visual representation, and actually collating them.

In this section and for the most part of this chapter, we have focused on specific data analysis tasks that are to be accomplished visually and interactively by a human user in one way or the other. However, we largely ignored the question regarding the technology used for interaction and visual output. We implicitly assumed to be working with a normal display, mouse, and keyboard. In the next section, we will leave these behind and study how interactive data analysis can be carried out in alternative display environments and with different interaction modalities.

4.8 INTERACTION BEYOND MOUSE AND KEYBOARD

What most of the existing visualization and interaction approaches have in common is that they are targeted for regular desktop workplaces. Yet, already in 1985, researchers recognized the importance of considering new technologies for the interface between humans and computers [HHN85]:

"But if we restrict ourselves to only building an interface that allows us to do things we can already do and to think in ways we already think, we will miss the most exciting potential of new technology: to provide new ways to think of and to interact with a domain."

Hutchins et al., 1985

EGA, MCGA, and VGA graphics cards and the computer mouse were the new technologies at that time. Today, new display technologies are related to large display walls or small mobile displays, both with high-resolution graphical output. New form factors and the increased pixel density made it necessary to adapt existing visualization approaches or to devise new ones. Modern interaction technologies, such as multi-touch interaction or tangible interaction have considerably broadened the spectrum of what is possible and, at the same time, created a need for rethinking existing visualization solutions with regard to interaction.

In this section, we will shed some light on how interactive visualization solutions can work with modern input modalities and output environments. We will start with basic touch interaction. From there, we continue with interaction for tangible visualization views. Finally, we illustrate how proxemic interaction can facilitate visualization on a large high-resolution display wall.

4.8.1 Touching Visualizations

Touch interaction has become very popular, especially for hand-held devices. By touching the visualization directly under our fingertips, also the exploration and analysis of data seems to be particularly well supported and promising. With touch interaction, direct manipulation becomes truly direct because the interaction takes place where the visualization is shown: on the display.

Yet, in order to actually obtain directness, we have to address a few challenges. First, we have to deal with the fact that only two states of Buxton's three-state model of graphical input (see Section 4.3.1) are available when using current touch technology. That is, there is no way we can hover a visualization as we can do with a mouse. We can either touch it or not.

Second, touch interaction is less precise. While practiced users can position a mouse pointer at a pixel-precise location, we are typically unable to touch down exactly at a certain pixel. This is due to our limited motor skills and the fact that our fingertips are several times larger than a pixel. Moreover, as the hover state is lacking, there is no way of correcting the initial touch point before triggering an action. Once the finger is down on the display surface, the touch is registered.

Another problem is the occlusion that is introduced when parts of our hand or arm cover information on the display. Moreover, dealing with multiple touches is non-trivial from an interaction design perspective. An in-depth discussion of all of the aforementioned challenges of touch interaction for visualization is beyond the scope of this chapter. For more details, the interested reader is referred to the further readings collected at the end of this chapter.

Here, we would rather briefly demonstrate the differences of mouse-based and touch-based interaction for a simple visual representation of time-oriented data, a SpiraClock visualization.

Touch-enabled SpiraClock Visualization

The SpiraClock is a technique for visualizing collections of temporal events, such as the personal agenda or bus schedules [DH02b]. The SpiraClock's hands display the current time, whereas its interior provides a spiral view onto the time to come. Along the spiral, future events are marked as spiral segments, each spiral cycle represents an hour in the future.

Figure 4.44 shows a simple example with the current time being almost 3:55 o'clock. In 5 minutes, the next appointment will start and it will take 15 minutes. Following the spiral inward, we see that after a half-hour break there will be another appointment lasting 25 minutes. Still further inward we see more future events. As time goes by, events gradually move outward and eventually exit the spiral, while future events enter the spiral at the center.

In order to enable users to explore events in time and adjust the SpiraClock, it makes sense to support the following basic interactions:

- Navigate in time

- Adjust future view

- Query details

The question that interests us is how these actions can be carried out with the classic mouse and alternatively with touch interaction. Let us first look at the mouse-based approach and then at the necessary changes to create a touch-enabled SpiraClock.

Figure 4.44 Visualizing future appointments with a SpiraClock.

Mouse Interaction For navigation in time, users can temporarily set a different time by rotating the clock hands via drag gestures. The view into the future can be narrowed or widened by dragging the spiral. Dragging toward the center will reduce the number of cycles, dragging outward to the rim will bring more cycles to the SpiraClock's interior. Finally, textual details can be queried simply by hovering an appointment, which will display a tooltip with the corresponding information.

Touch Interaction In order to take advantage of touch interaction, the mouse-based design has to be cast into a touch-based one. On first sight, the transition from mouse to touch seems trivial: Instead of operating the mouse, fingers will carry out the drag gestures directly on the display. Yet, our users may soon stumble upon the difficulty of grabbing the clock hands due to the limited touch precision.

One way to address the precision issue would be to enlarge the clock hands. Yet, thicker clock hands not only look awkward, they also increase occlusion. An alternative is to decouple graphics and interaction [Con+08]. That is, regular clock hands are used for drawing the SpiraClock and a separate, slightly thicker invisible geometry for handling the interaction. This way, the SpiraClock can maintain its appearance, while the clock hands are easier to pick thanks to their invisibly increased thickness.

Unfortunately, the enlarged interaction geometry of the clock hands makes it more difficult to perform a drag gesture on the clock interior, as we might accidentally touch the hands. To solve this conflict, we can use a dedicated

TABLE 4.4   Mouse-based vs. touch-based interaction for the SpiraClock

Action	Mouse		Touch	
Navigate in time		Drag gesture on clock hands		Drag gesture on *enlarged* clock hands
Adjust future view		Drag gesture on spiral interior		*Pinch gesture anywhere*
Query details		Hover appointments		*Tap on* appointments

touch gesture: the pinch gesture. While clock hands are rotated with a single finger, manipulations of the spiral will be performed by a two-finger pinch gesture. Pinching is the quasi-standard for scaling interactions on touch devices. In our case, we scale the number of spiral cycles as soon as a second finger touches the display.

Finally, we need to design the interaction to query details about the displayed events. As hovering is incompatible with touch interaction, we have to resort to a different approach. This time, we use a simple tap gesture, which corresponds to just a brief touch. When an event is tapped, a tooltip will display its details. A second tap on the same event (or on the background) will dismiss the tooltip.

But what about the increased thickness of the clock hands? Doesn't it lead to problems when the clock hands overlap with events? Yes, indeed. But the tap gesture is a discrete interaction, whereas the dragging of clock hands is continuous. This means, the tap is a very short interaction that is easily detected. Once we know that a tap occurred, we simply ignore the clock hands and consider only the events as potential interaction targets.

As we have seen, already for the three simple interactions of the SpiraClock, we had to carefully think about the transition from mouse to touch interaction. Table 4.4 provides a comparison of both designs. We cannot claim that either of the designs is better or even optimal. Yet, they served well the purpose of demonstrating a few issues when designing touch interaction for visual representations of data.

In the next section, we will continue with modern ways of exploring data. Yet, we will go one step further from fingers touching visualizations *on* a display to tangible interaction *with* the display.

4.8.2   Interacting with Tangibles

Tangible user interfaces is a broad field of research [SH10]. The goal of tangible interaction is to narrow the gap between the virtual world on the computer and the real world in which the interaction takes place. The link between both worlds is physical objects, so-called *tangibles*. Physical manipulations of tangibles are transferred to virtual objects on the computer. In this sense, mouse-based interaction is already a form of tangible interaction, yet a rather indirect one.

More direct tangible interaction can be achieved by using tangibles such as discs or cubes directly on horizontal touch-sensitive surface displays. To minimize occlusion, the tangibles are typically made from translucent materials, such as acrylic glass or foil. Various physical gestures can be performed with tangibles. Tangibles can be placed and moved across the display. Additionally, they can be rotated or positioned with different sides facing upwards. These tangible interactions expand the possibilities for interactive data exploration on touch-sensitive displays. Yet, interaction and visualizations remain in the two-dimensional horizontal plane of the display.

Next, we present an approach that extends tangible interaction to the three-dimensional space above the display to provide enhanced visualization and interaction functionality.

Tangible Visualization Views

Let us first introduce the basic setup. As before, a horizontal surface display will serve to visualize data and to receive touch input from the user. What we add to this basic setup are so-called *tangible views* [Spi+10].

Tangible views are lightweight "devices" that act as additional displays in the space on or above the horizontal surface. In a most simple instantiation, a tangible view can be a piece of cardboard onto which a projector transmits visualization content. In this case, the tangible view is passive. A tangible view can also be active, in which case it is capable of displaying graphical content on its own, for example, a tablet device.

A key characteristic of tangible views is that they are spatially aware. Through constant tracking, the system always knows a tangible view's position and orientation. This opens up whole new possibilities for interaction as illustrated in Figure 4.45. The extended capabilities of tangible views include basic translation and rotation in three dimensions as well as gestures of flipping, tilting, and shaking. By providing tangible views that are distinguishable by shape or appearance it is possible to create an interaction toolbox, where users can infer interaction functionality from the look of a tangible view. Multiple tangible views can be used simultaneously for advanced interaction and for adding display space for visualization purposes.

Yet, offering extended interaction capabilities is only one part of the story. The second part is to utilize them to create a semantically meaningful interaction vocabulary for visualization scenarios. This has to be done depending on the characteristics of the data and their visual representation, and in line with the tasks to be carried out.

Next, we take a closer look at two selected applications of tangible views. In the first example, we will use a tangible view as a magnifying lens for a scatterplot visualization. For the second example, we apply two tangible views

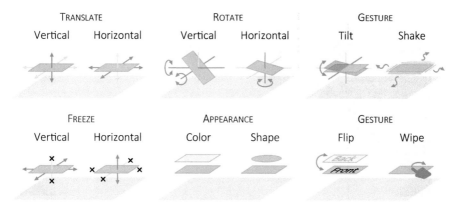

TRANSLATE | ROTATE | GESTURE
Vertical   Horizontal | Vertical   Horizontal | Tilt   Shake

FREEZE | APPEARANCE | GESTURE
Vertical   Horizontal | Color   Shape | Flip   Wipe

Figure 4.45   Extended interaction with tangible views. Adapted from [Spi+10].

to visually compare graph matrices. For both examples, we have already seen non-tangible implementations earlier in this chapter. It will now be interesting to see how tangible interaction creates a more physical data exploration experience.

A Tangible Magnifying Lens

For our first example, the main visualization is a scatter plot. In order to de-clutter dense parts of the plot, we can use a magnifying lens as described in Section 4.6.3. In a standard mouse-plus-keyboard setting, the lens can be moved across the visualization by drag gestures to define where it should take effect. The degree of magnification is typically adjustable via standard or custom-made sliders.

Now let's make the virtual lens tangible and truly direct. To this end, a circular tangible view is inserted into the space above the surface display as illustrated in Figure 4.46. The tangible view's horizontal position determines which part of the visualization is to be magnified (dashed circle on the surface). The actual lens effect is projected onto the tangible view. This already yields a very tight coupling of display and interaction.

The extended interaction vocabulary also makes it possible to control the magnification factor in a tangible way. There are several options how this can be implemented. The tangible view can be raised and lowered along the vertical axis to increase and decrease magnification. Figure 4.46 illustrates a second alternative: The tangible view is rotated around the vertical axis while a circular gauge visualizes the current magnification factor.

As a result, we obtain a tangible magnifying lens that can be adjusted by physical manipulation, and the visual feedback is immediately visible where the interaction takes place. In this first example, one or two hands control a single tangible view. Next, we add a second tangible view and operate both simultaneously to accomplish comparison tasks.

Figure 4.46 A circular tangible lens for magnification purposes.

Figure 4.47 Comparing matrix data with two tangible views.

Tangible Visual Comparison

Now, the base visualization on the surface shows a graph. It is represented in matrix form as described in Section 3.5.2 of Chapter 3. We will be using two rectangular tangible views as depicted in Figure 4.47. In order to select a sub-matrix to be compared, the user moves a tangible view horizontally above the surface. To fix the selection, a freeze gesture is carried out. This involves no more than swiping the thumb through a particular spot at the border of the tangible view.

Once frozen, a tangible view maintains its visualization content, effectively ignoring any horizontal movements. This allows the user to arrange both tangible views side-by-side for closer inspection and comparison. Once the views are sufficiently close to each other, the system recognizes the user's comparison intent and adds an explicit encoding of the overall aggregated similarity of the matrices to the tangible views. In our case, the green frame suggests that the matrices are quite similar to each other.

 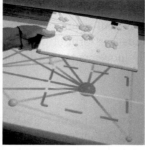 

(a) Parallel coordinates.   (b) Node-link diagram.   (c) Space-time cube.

Figure 4.48   Tangible views for different visualizations.

In order to continue the visual comparison elsewhere in the data, the tangible views need to be awakened from their frozen state. A natural gesture to stir up a stationary system is to shake it. This is what the user does to un-freeze the tangible views: Simply shake them horizontally and start over with selecting different parts of the matrix.

In summary we see that tangible views offer novel and interesting ways of interacting with visual representations of data. This is not only the case for magnification and comparison tasks, but for a broad range of visualization problems. Figure 4.48 illustrates further applications examples. In Figure 4.48a, the tangible view can be raised or lowered to control the sampling rate with which the multivariate parallel coordinates plot is sampled. Raising and lowering the tangible view above the node-link diagram in Figure 4.48b enables the user to access different levels of abstraction of a hierarchical graph. The tangible view in Figure 4.48c serves as a slice in a space-time cube visualization of spatio-temporal data.

From a conceptual perspective, tangible views advance interaction and visualization in several ways. Tangible views integrate display and interaction device, allowing us to interact directly *with* the visualization. They also extend common two-dimensional interaction by tangible three-dimensional interaction above a horizontal base visualization. The resulting enhanced interaction and the extended physical display space create a tangible experience of otherwise purely virtual actions.

Results of controlled user studies suggest that tangible spatial interaction is indeed a promising alternative when working in and with layered zoomable information spaces, which are common in visualization scenarios [SMD12; Spi+14]. Yet, developing an enhanced interaction and providing evidence of its extended expressive power is only a beginning. It remains to be investigated what will be the most suitable interaction designs for a broad range of data analysis scenarios, taking into account different types of data and different user tasks.

Another interesting observation is that tangible views, touch-sensitive devices, and the classic mouse, all require us to use our hands for interaction. Next, we will look at a hands-free form of interaction where a visualization is controlled by physical body movements in front of a large display wall.

4.8.3 Moving the Body to Explore Visualizations

So far, we have considered visualization on regular displays, horizontal surfaces, and tangible views. These and other output devices with conventional pixel resolution are typically limited in the amount of information that can be displayed. Thanks to technological advances, large high-resolution displays are now becoming available to a broader range of users. The larger physical size and the increased pixel resolution offered by such displays have obvious advantages for visualization applications. Particularly in light of big data, being able to visualize much more information is an exciting prospect.

Yet, with larger size and more pixels there also come new challenges. It is no longer feasible to interact with a mouse or touch alone, because it is simply too difficult, if not impossible to reach across the entire display. Therefore, new solutions are needed to support the visualization and interaction on large high-resolution displays. In this section, we look at a scenario where physical body movements in front of a display wall support the exploration of a graph with multiple levels of detail [Leh+11].

Visualizing a Graph on a Display Wall

The setup we are addressing is a tiled display wall as shown in Figure 4.49. The wall consists of 24 individual displays covering an area of 3.7 m × 2.0 m with a total resolution of 11,520 × 4,800 which amounts to 55 million pixels. The data we want to explore is a graph that is described at three levels of detail. The visualization is based on a node-link representation which is augmented with textual labels and hulls for grouping. Individual nodes can be expanded or collapsed in order to get a finer or coarser view on the graph, respectively. On a desktop, expand and collapse operations are typically carried out by clicking or tapping. In our scenario, this is obviously impractical due to the large distances that would need to be covered.

Interacting by Body Movements

An alternative is to control the displayed level of detail by the user's physical movement in front of the large display wall. This requires a tracking system to be set up to acquire information about the user's position and orientation (6 degrees of freedom).

The user's position can then be exploited to adjust the level of detail globally. To this end, the space in front of the display wall is sub-divided into zones with increasing distance to the display, as illustrated in Figure 4.50a. Each

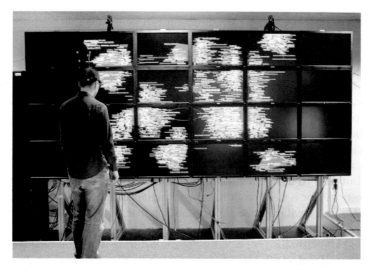

Figure 4.49 Graph exploration on a large high-resolution display wall. Reprinted from [Tom15].

zone corresponds to a level of detail. When the user moves into a zone closer to the display, the graph is visualized at greater detail. Stepping backward into zones farther away from the display will lead to a coarser representation. This approach is also dubbed *proxemic interaction* [BMG10]. It is inspired by natural human behavior: Humans typically step up to the object of interest to study it in detail and step back to obtain an overview.

With the zone-based interaction, users can control the level of detail globally. Local adjustments require a way to point at where the level of detail should be changed. One option to do this is to show more details exactly where the user is looking on the display. Based on the tracking information (position

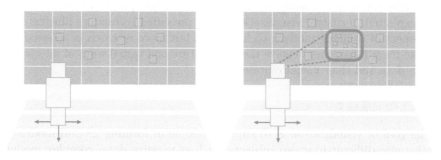

(a) Zones for global control. (b) Gaze plus lens for local control.

Figure 4.50 Interacting by physical movements. Adapted from [Leh+11].

and orientation), the user's gaze direction can be estimated. With the help of dedicated eye-tracking, the precision of the estimation can further be improved. At the spot being looked at, an interactive lens is embedded into the regular graph visualization as illustrated in Figure 4.50b. Nodes that are inside the lens are automatically expanded to reveal more detailed information. By moving the head, the user can quickly scan the graph for details. Filtering the tracking input and smoothly animating node expand and collapse operations help to avoid flickering caused by natural head tremor and to maintain a reasonably stable visualization.

The user's physical movements can be used not only to control the level of detail, but also to derive a suitable layout for the labels shown in the graph visualization. For a user standing close to the display, the visualization will show more and smaller labels. When looking from a greater distance, the user will see fewer, but larger labels for groups of nodes. This way, the costs for producing the visualization and for comprehending it can be balanced.

The results of a pilot study suggest that interaction through physical movement in front of a display wall can indeed be a valuable alternative in cases where classic means of interaction fail [Leh+11]. Physical movement not only better matches the scale of the display, it also is in line with natural interaction with real-world objects. Interacting via zones was reported as the approach that is easier to use, but on the other hand, the lens offered more control over where increased detail is to be shown.

Here, we considered the relatively simple task of adjusting the level of detail of a graph visualization. But physical movement has also proved useful in other scenarios. Zooming tasks, for example, can be supported well by physical navigation [Jak+13]. Other research results indicate that physical navigation can also be beneficial for higher-level analytic sensemaking [AN13].

In this last section of the chapter on interaction, we have focused on new technologies for interactive visual data analysis. We made a transition from mouse interaction to touch interaction and enhanced the latter with tangible interaction. Finally, we have illustrated how physical movements can be exploited to control visualizations. The presented techniques illustrate the prospect of utilizing modern technologies, but more research is necessary to make them reliable and evolve them to mature and tested ways of interacting with data and their visual representations.

4.9 SUMMARY

The goal of visualization is to support people in forming mental models of otherwise difficult-to-grasp subjects, such as massive data, complex models, or dynamic systems. The term *forming* implies that visual output is not the end product of visualization. It is rather the human-in-the-loop process of interactively exploring the data and adjusting their visual representation that enables us to gain insight. This chapter elaborated on this very process.

We looked at the topic from different angles, including aspects of the human user who interacts, the interaction tasks, the data being interacted with, and the technology used for interacting. In the first part of this chapter, we dealt with the question of *why* interaction is needed in visual data analysis scenarios. In the middle part, we discussed *what* users can actually accomplish with general interaction concepts and concrete interaction techniques. Finally, in the last part, we considered *how* interaction can be carried out using modern interaction modalities. Throughout the chapter, many examples illustrated how interaction can support data analysis activities.

Concluding Remarks

Useful and usable interaction techniques are the result of careful consideration of the human user, the analytic tasks to be accomplished, the characteristics of the data, and the technological environment in which the analysis takes place. Several books would be needed to cover the design space for interaction comprehensively. Nonetheless, the following paragraphs will provide some high-level remarks on the design of interaction for visual data analysis.

Duality of Input and Output   A fundamental requirement is to consider the duality of visual output and interactive input right from the beginning. Hutchins and colleagues put it this way [HHN85]:

"[...] the nature of the relationship between input and output language must be such that an output expression can serve as a component of an input expression."

Hutchins et al., 1985

Applied to the context of this book, this means that whatever we display in a visualization will most certainly be relevant for interaction as well. Whatever we desire to input into the system will typically require a suitable visual representation for direct manipulation. Adding interaction as an afterthought to a visualization is likely to cause trouble and increased development costs.

Mapping Tasks to Interactions   Given the complexity of exploratory and analytic activities and the wealth of interaction modalities, finding a good mapping of tasks to concrete interactions is a non-trivial endeavor. Two requirements are particularly important: Interactions must be conflict-free and should be cost-efficient.

It is absolutely necessary that interactions be *conflict-free*, meaning that an interaction must be associated with a unique task. To this end, interactions must be unambiguously distinguishable. A distinction is typically made at the level of gestures, such as hovering, clicking or double-clicking. Also the space where the interaction takes place can be a differentiating factor. That is, we use the same gesture, but perform them in distinct spatial regions. In this case, clicking a map, clicking a data item, and clicking the background are all different interactions. Moreover, the order or timing of operations could be considered, but this complicates the interaction enormously.

Speaking of complicated interaction, the second requirement is about interaction costs: Interaction should be *cost-efficient*. Each interaction costs. The more often an interaction is used, the higher the accumulated costs. Therefore, the mapping of tasks to interactions should consider a ranking of the relevance or frequency of tasks and an estimation of the costs of the available interactions. Based on that, a reasonably balanced mapping can be established. Frequent tasks should be mapped to interactions with the lowest costs. Only for infrequent tasks is it acceptable to use interactions with higher costs.

Interactive plus Automatic   Directly related to the cost aspect of interaction is the question regarding what should be done interactively and what can be accomplished by automatic means. But this is not a question of interactive *versus* automatic means, rather it is a call for interactive *plus* automatic means. This chapter proposed several ideas of how interaction can be eased by integrating automatic mechanisms. After all, it is the responsibility of the system (and the designer of the system beforehand) to provide the information needed in a particular situation. Whenever we make functionality available via interaction we should carefully think about how interaction costs can be reduced by integrating automatic assistance.

With these concluding remarks, we close the chapter on interaction. As we have seen, interaction is an integral component of visual data analysis approaches. In the next chapter, we will see that analyzing large data additionally requires the integration of computational analytic components.

FURTHER READING

General Literature: [Dix+04] • [SP09] • [Tom15] • [SP16] • [DP20]

Zoomable Visualizations: [BH94] • [FB95] • [Fur97] • [Bed11]

Lenses for Visualization: [Bie+93] • [TFS08a] • [Tom+17]

Visual Comparison: [Gle+11] • [vLan18] • [Gle18]

Beyond Mouse and Keyboard: [Lee+12] • [Ise+13] • [JD13] • [Mar+18]

Automatic Analysis Support

CONTENTS

I N THE PREVIOUS chapters we described fundamental approaches to interactive visual data analysis. In light of the potentially complex and very large datasets we are facing today, it is hardly possible to indiscriminately visualize and interact with all data. Our visual representations would simply be overcrowded and interaction would be cumbersome and costly. This is why computational analysis support is necessary. Visual analytics pioneer Daniel Keim puts it this way [Kei+06]:

"The visual analytics process comprises the application of automatic analysis methods before and after the interactive visual representation is used. This is primarily due to the fact that current and especially future data sets are complex on the one hand and too large to be visualized in a straightforward manner on the other hand."

Keim et al., 2006

The use of automatic analysis methods aims at extracting essential data characteristics. Showing key characteristics instead of the original data values reduces the complexity of visual representations and facilitates an initial overview of the data. Deeper insight can then be achieved by combining visual analysis, interactive querying, and further automatic computations. This procedure is nicely reflected in the visual analytics mantra, which we already mentioned in the introduction [Kei+06]:

"Analyse First –
Show the Important –
Zoom, Filter and Analyse Further –
Details on Demand"

Keim et al., 2006

The primary goal of this chapter is to provide an overview of computational approaches that can support the analysis of large and complex data. While we will focus on the *analyze* step, we will also see that it is actually the tight interplay of automatic, visual, and interactive means that really advances the way data can be investigated and understood.

Each section of this chapter will briefly explain a basic strategy for the analysis step and one or two selected techniques that implement the strategy. A common theme will be the *reduction of complexity* to make the visual analysis easier or enable it at all. What gets reduced will differ from section to section. In Section 5.1, our objective will be to reduce the complexity in visual representations. Section 5.2 considers the extraction of data and features of interest to narrow down the analysis on the parts that are relevant.

How complexity can be reduced in the data space will be described in Sections 5.3 to 5.5. Section 5.3 is dedicated to the reduction of the cardinality of the data domain via data abstraction methods. In Section 5.4, we will see that grouping similar data is a powerful strategy to reduce the number of data elements. Finally, Section 5.5 will explain dimensionality reduction as a way to focus the analysis on the key information-bearing data variables.

Let us now discuss the sketched options in more depth. We will start with methods that aim at reducing the complexity in visual representations.

5.1 DECLUTTERING VISUAL REPRESENTATIONS

Over-plotting and visual clutter are common problems when large volumes of data are visualized. A first and important strategy facilitating the analysis of large data is to employ methods that can declutter the visual representation. We consider two basic approaches: computing and visualizing density as well as bundling of geometrical primitives.

5.1.1 Computing and Visualizing Density

Density-based representations aim to communicate the distribution of data, rather than showing individual data values. The basic idea is to calculate how many data values fall within certain intervals. Alternatively, one can calculate how many graphical objects are within certain regions of the display. Both alternatives will next be illustrated by two well-known examples: continuous scatter plots, which are based on *data density*, and outlier-preserving focus+context visualization, which is based on *visual density*.

Data Density   Traditional scatter plots visualize data elements as dots. For very large data, this can result in severe over-plotting, which makes it impossible to discern how much data are represented by a dot.

Continuous scatter plots solve this problem by visualizing a continuous density function [BW08a]. For this purpose, a mapping is performed from the data domain to the spatial domain that is spanned by the two axes of the scatter plot. While being conceptually continuous, the density is typically

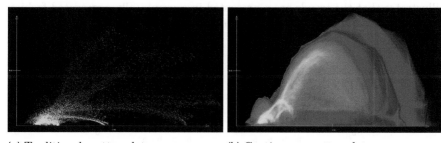

(a) Traditional scatter plot.   (b) Continuous scatter plot.

Figure 5.1   Comparison of traditional scatter plot and continuous scatter plot. Reprinted from [BW08a].

approximated by counting data elements in discrete intervals. The continuous visualization is then obtained by interpolation.

Figure 5.1 compares a traditional scatter plot based on discrete dots with a continuous scatter plot based on data density. Both plots show the same "blunt-fin" dataset. With the density-based visualization, internal structures in the data are much easier to discern.

Visual Density In contrast to scatter plots, parallel coordinate plots visualize data elements as polylines across parallel axes. For large datasets, a large number of polylines needs to be drawn, which can substantially clutter the visual representation. Again, a density-based approach can be employed to reduce visual clutter. Yet, now we are interested in the visual density of the graphics [NH06]. By computing the visual density, it is possible to accentuate not only general trends, but also details such as outliers. The rest of the data can be attenuated.

Figure 5.2 schematically depicts how the visual density in parallel coordinates can be determined and how the visualization can be adapted accordingly. For the purpose of illustration, we start out in Figure 5.2a with the most basic (and unusual) case of only two parallel axes.

In a first step, the axes are subdivided into bins $b_{i,j}$, where i denotes the axis and j the bin per axis. The second step is to count how many lines emanate from the bins of one axis and arrive at the bins of the other axis. These pairwise frequencies are stored in a so-called *bin map*. We can see in Figure 5.2b that a bin map is a grid of cells, where each cell represents a pair of bins. For example, there are four lines between bins $b_{1,1}$ and $b_{2,2}$. Consequently, the corresponding cell in the bin map contains the number 4. Repeating the counting for all pairs of bins completes the bin map.

The third step is to categorize the cells of the bin map as belonging to trends or details, which is illustrated in Figure 5.2c. Thresholds, outlier detection, and clustering methods can be involved in the categorization [NH06]. In the final step, trends and details are drawn to the display, but differently so. Figure 5.2d illustrates that the trends are represented in a graphically aggregated fashion, whereas the details get rendered as individual lines.

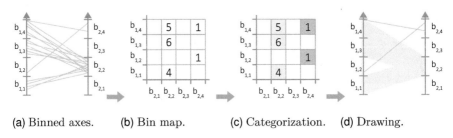

(a) Binned axes.   (b) Bin map.   (c) Categorization.   (d) Drawing.

Figure 5.2 Procedure of determining visual density in parallel coordinates.

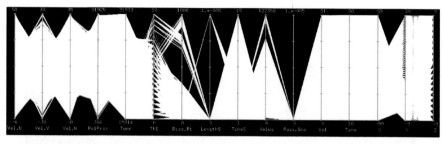

(a) Standard parallel coordinates plot.

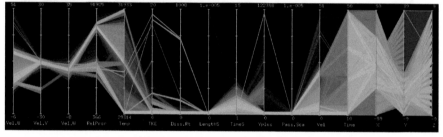

(b) Density-based visualization.

Figure 5.3 Visualization of more than three million data elements. Reprinted from [NH06].

For general parallel coordinates plots with m axes, the above procedure is repeated for each pair of neighboring axes. As a result, $m - 1$ bin maps are obtained based on which the visualization complexity is reduced globally.

Figure 5.3 demonstrates the benefit of the described approach. A classic parallel coordinates plot is shown in Figure 5.3a. Due to clutter, it does not reveal much of the data's inner structure. The density-based approach is shown in Figure 5.3b. The general trends are visualized as parallelograms of different shades of green, where strong trends use brighter greens and weaker trends use darker greens. Data elements that have been categorized as details are explicitly visualized as green polylines. With the density-based visualization, trends and details can be seen much better.

By interactively adjusting the number of bins and the thresholds for categorizing trends and details, it is possible to further abstract or elaborate on the visualized structures. Moreover, red polylines can be superimposed on the visualization to facilitate comparing user-selected data against the automatically determined trends. This is illustrated in Figure 5.4.

We have seen that density-based methods, no matter if they operate on data values or on graphical objects, are very well suited to declutter visual representations. In the next section, we discuss another approach that addresses the problem of visual clutter, this time, by bundling geometrical primitives.

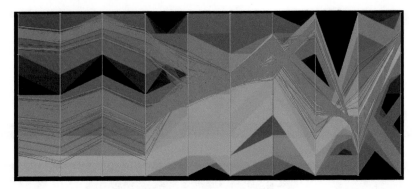

Figure 5.4 User-selected data in red compared against general trends in green. Courtesy of Helwig Hauser.

5.1.2 Bundling Geometrical Primitives

As we know from the previous section, visual representations that use lines or line segments as their basic geometrical primitives are prone to severe clutter. Many such visualizations exit, for example, the parallel coordinates plot as described before, node-link diagrams of graphs, or trails of movement data. In this section, we focus on decluttering such visual representations by summarizing geometrical line primitives into so-called *bundles*. As before, the goal is to highlight the fundamental structures in the visual representation.

The general procedure of bundling is outlined in Figure 5.5. The starting point is line primitives. They are the basis for the specification of paths that consist of the lines' original vertices plus additional control points. Such paths can be flexibly adjusted in their course in order to form visual bundles. To this end, it is necessary to decide which paths should belong to a bundle and how the individual paths should be transformed. Both decisions can be made based on either an explicit or an implicit bundling definition.

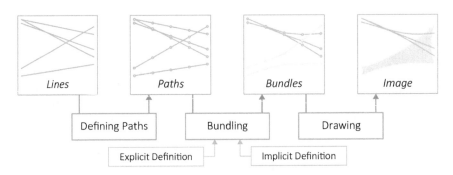

Figure 5.5 General procedure of bundling. Inspired by [LHT17].

(a) Conventional representation.   (b) Hierarchical edge bundling [Hol06].

Figure 5.6   Visualization of dependencies in a software class hierarchy. Adapted from `bl.ocks.org/mbostock/4341134`.

Explicit definitions resort to existing criteria. For example, paths can be bundled along given hierarchical structures [Hol06]. In parallel coordinate plots, polylines can be bundled according to pre-calculated clusters [Hei+12]. Bundling based on such predefined criteria produces predictable layouts, but it is less flexible in terms of bundling control.

Implicit definitions make use of appropriate similarity metrics. Paths that are similar will be transformed to bundles. A prominent example is force-directed bundling methods [HvW09]. Compatibility criteria define which paths are eligible for bundling and the actual bundling transformation is done via spring forces between the control points of compatible paths. The implicit bundling strategy is more flexible, since it is not necessary to address pre-defined constraints. It can be adjusted easily, and bundles will be updated on the fly.

In the last step, the bundles are visualized. There are two options: drawing individual curves or drawing entire bundles as compact geometrical shapes. The important thing is to ensure that bundles are recognizable and distinguishable. This can be achieved by a suitable visual encoding. Often, bundles are associated with a unique color. The visual encoding can be enhanced by shading the bundles based on a pseudo-surface that spans a bundle [LHT17]. Blending can be applied to resolve occlusion problems when bundles overlap.

Figure 5.6 illustrates the positive effect of bundling. The treemap in the background visualizes the class hierarchy of a software framework. Links between the nodes of the treemap indicate dependencies between classes. The darker a link is, the larger are the classes involved in a dependency. As can be seen in Figure 5.6a, many links lead to substantial clutter. By transforming the links into curved bundles, in this case based on the explicitly given hierarchy, the visualization can be streamlined. Figure 5.6b reveals the major relations among the class dependencies.

In summary, this section discussed methods to facilitate the analysis of large datasets by reducing the complexity in visual representations. In the following sections, we will continue with reducing the complexity in the data.

5.2 FOCUSING ON RELEVANT DATA

A common and widely used strategy in terms of data reduction is to focus the visual analysis on data of interest. But what are the data of interest and how can a distinction between interesting and less-relevant data be made? Of course, users could interactively select the data they deem important. For large datasets, however, a purely manual procedure can be time-consuming and error-prone. How automatic computations can help in determining relevant data will be the topic of this section.

The basic idea is to let users specify (not select) their interests and to employ automatic computations to select those parts of the data that match the specification. We discuss two concepts that implement this general idea. First, we study the concept of degree of interest and how it can help us to narrow down the analysis on relevant data. Second, we present the concept of feature-based visual analysis, which is about the specification and automatic extraction of meaningful data characteristics.

5.2.1 Degree of Interest

The *degree of interest* (DoI) is a concept to capture the relevance of data for solving an analysis task. The degree of interest can be expressed by *DoI functions*. A DoI function assigns to each data element a relevance value. Using a suitable threshold, it is then possible to distinguish between relevant and less-relevant data. The relevant data correspond to the target of an analysis task. In Section 2.2.2, we explained that particularly the targeted data need to be visualized faithfully. In contrast to that, less-relevant data can be dimmed or even omitted. This substantially reduces the complexity of the visual analysis.

Basic Approach

Already in the 1980s, Furnas introduced a DoI function to express the degree of interest for the nodes of static hierarchies [Fur86]. The basic idea is that the degree of interest of a node n_i depends on the distance of n_i to a focus node n_f and an a priori interest of n_i. For the purpose of illustration, we set up a DoI function $doi(n_i, n_f)$ as follows:

$$doi(n_i, n_f) = dist(n_i, n_f) + api(n_i)$$

The focus node n_f is assumed to be in the user's center of attention, $dist(n_i, n_f)$ is the length of the shortest path between n_i and n_f, and $api(n_i)$ describes the a priori importance of n_i. In the context of hierarchical data, it makes sense to define $api(n_i)$ as the level at which n_i can be found in the hierarchy, that is, the distance of n_i to the hierarchy's root.

Figure 5.7 illustrates how the DoI function assigns a relevance value to each node. In Figure 5.7a, we can see the $dist(n_i, n_f)$ term, to which the $api(n_i)$ term is added in Figure 5.7b. As our DoI function is based on distances,

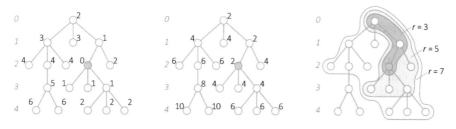

(a) Distances to focus node. (b) Adding node levels.　(c) Extracted subtrees.

Figure 5.7 Illustration of Furnas' DoI function.

lower values stand for higher degrees of interest. A user-defined threshold r determines where to draw the line between relevant and non-relevant nodes. Depending on the threshold used, relevant sub-structures of different size are automatically extracted from the hierarchy. Figure 5.7c provides examples for three different thresholds $r = 3$, $r = 5$, and $r = 7$.

Furnas' idea of a DoI function is well established. It has been extended in different ways, for example, to handle multiple focus nodes [HC04] or to deal with general graphs [vHP09]. Most of the time, however, DoI functions are pre-defined and cannot be altered by the user. This is in contrast to the necessity of adapting the visual analysis to changing tasks and requirements. Next, we look into a more flexible approach that allows the user to construct DoI functions on demand.

Flexible Construction of DoI Functions

The construction of a DoI function requires a modular design of the involved components [Abe+14]. Without going into too much detail about the underlying formalism, setting up a DoI function can be done in three steps:

1. Definition of relevance components

2. Combination of relevance components

3. Specification of relevance propagation

With the first step, the user defines *relevance components* that compute a relevance value per data element. It makes sense to work with normalized values, where 0 signifies no relevance and 1 is the highest relevance. An example of a relevance component would be a Gaussian function that produces a smooth decline of relevance with increasing distance to a particular value of interest. It can make sense to define several components to capture different aspects of relevance. For example, one component could consider spatial proximity, another component could account for temporal dependencies. Also structural properties can be taken into account, as in the previous example, where we used distances in a hierarchical structure.

The second step is to construct a comprehensive DoI function by combining the relevance components. Conceptually, a combination of relevance components can be modeled as a function that takes two or more relevance values and returns a combined relevance value. Furnas' DoI function simply adds the components. Yet, more advanced combinations are possible. A common example is the combination as a weighted sum. It allows for balancing the influence of components: Heavily weighted components influence the overall result more than lightly weighted ones. Furthermore, *min* or *max* combinations can be applied. A *min* combination requires all involved components to return high relevance in order for the combination to return a high relevance as well. A *max* combination works the other way around: If either component returns a high relevance, the overall combination will yield high relevance.

We can now flexibly construct component combinations to compute the relevance of data elements. However, a relevance value characterizes only an individual data element. The immediate context of the data elements is not taken into account. For example in Figure 5.7c, a subtree with three nodes is extracted for $r = 3$. Yet, structural properties of the focal node are lost. The subtree does not tell us whether the focal node has siblings or child nodes. Choosing a higher threshold of $r = 7$ adds too much context information than would be needed specifically for the focal node. Only if the threshold is set to $r = 5$ is the user's interest in siblings and child nodes properly reflected. However, finding such an appropriate threshold is notoriously difficult.

Therefore, the third and last step is to decide on a propagation method that distributes relevance in the vicinity of high-relevance data elements. The propagation can be performed in different ways. Structural propagation distributes relevance along the edges in a graph, temporal propagation considers neighboring time intervals, and spatial propagation spreads relevance in a spatial neighborhood. By specifying different types of propagation, the user can trigger the extraction of different context information.

Taken together, users can now specify what data should be considered as relevant. The flexible combination of relevance components makes it possible to specify which data aspects should be taken into account and how they should impact the final decision. With propagation, it is possible to steer how high relevance can diffuse into neighboring data.

So far, the described procedure provides but an abstract conceptual scheme. In order to make it applicable to real analysis problems, we need a suitable user interface that supports the three aforementioned steps. Such a user interface typically requires a design that is tailored to the data being studied. Next, we will consider this aspect in more depth by the example of an interface for DoI-based visual analysis of dynamic graphs.

DoI-based Visual Analysis of Dynamic Graphs

Dynamic graphs are typically large and complex, which makes their visual analysis difficult. Consider, for instance, dynamically changing co-author networks, where nodes represent authors, and edges exist between authors that published together. Node attributes provide additional information about the authors, for example, the number of publications. Edge attributes, such as the number of joint publications, characterize the co-authorships.

Figure 5.8 shows five years of such a dynamic co-author network as extracted from the DBLP computer science bibliography database. From the visualization, we can very clearly see how computer science has gained in importance as an academic discipline over the years. But we can also see that the network structure drowns in a flood of new authors and co-author relationships. Already for the year 1990, the visualization is so saturated that no structural information can be discerned from it.

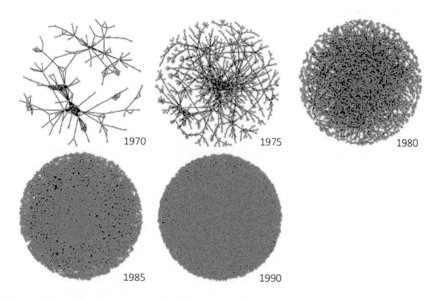

Figure 5.8   Five years of a dynamic co-author network extracted from DBLP. Courtesy of Steffen Hadlak.

In the following, we will demonstrate how the flexible DoI-based mechanism outlined before can help us in analyzing large dynamic graphs [Abe+14]. The graph that we will be dealing with covers 22 years of the DBLP database from 1990 to 2011 with an overall number of 914,492 nodes and 3,802,317 edges. The DoI-based visual analysis starts with the formulation of some interest in this big network.

Let's assume the user is interested in the top authors. Now, what makes an author a top author? Obviously the number of publications of an author plays a role. Moreover, top authors are likely to have many co-authors with whom

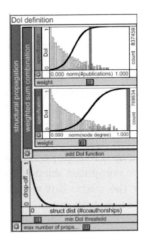

Figure 5.9 DoI specification via a nested graphical user interface. Reprinted from [Abe+14].

they published their many papers together. So, two relevance components need to be defined: one based on the number of publications, the other based on the number of co-authors, where we would like to prioritize the number of publications a little more. Finally, it must be decided how the relevance values should be propagated.

In order to enable users to specify interactively what we described verbally so far, a suitable graphical user interface is needed. Figure 5.9 shows an interface that is based on nested frames. The innermost frames represent our two relevance components. They are designed as histograms. Sigmoid functions map high numbers of publications and co-authors to high relevance values. The inner components are encompassed in an intermediate frame representing a weighted sum. Sliders at the components facilitate the adjustments of weights to give the number of publications a slightly higher impact. Finally, the outermost frame represents the selected propagation method, in our case a structural propagation with a certain drop-off. Thanks to the tight integration of aggregated visual representations and user controls, the described interface makes it easy to extend and adjust the DoI specification as necessary.

The next question is how the DoI approach can help us reduce complexity and support the visual analysis of our large dynamic graph? In Figure 5.10, we can see that the DoI interface is but a part of a larger framework. In addition to the DoI interface (a), there are a panel with graph statistics (b), a list of all authors and their co-author relationsships for manual selection (c), controls to adjust visualization thresholds (d), a time line with the aggregated relevance per year (e), and a node-link representation of the graph (f).

The node-link representation visualizes the graph in a DoI-based fashion. Nodes of interest, that is, nodes with a relevance above the user-defined threshold are represented in full detail as individual dots. Subgraphs that are

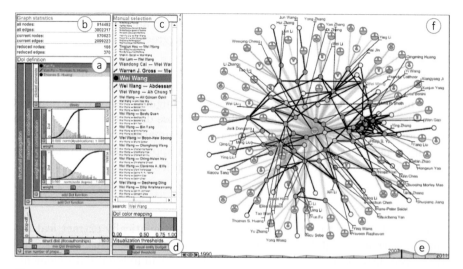

Figure 5.10   DoI-based visualization of the 2007 DBLP co-author network. The graphical interface controls (a)–(d) allow users to specify their interest in the data. The visualizations (e) and (f) show the relevant data and their associated relevance values. Reprinted from [Abe+14].

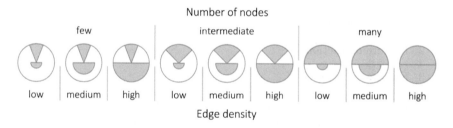

Figure 5.11   Glyph design for representing collapsed subgraphs. Adapted from [Abe+14].

not of interest are collapsed and summarized by glyphs. The glyphs visualize two properties of the collapsed subgraphs: their number of nodes and their edge density. As illustrated in Figure 5.11, the number of nodes is encoded by the angle of the arc sector at the top of a glyph, whereas the edge density is mapped to the radius of the semicircle at the bottom. Throughout the node-link representation, the relevance values are encoded with variations of green, where darker green stands for higher relevance. Glyphs and edges leading to glyphs are shown in gray to indicate that they are less relevant.

Comparing the representation of the network of 1990 in Figure 5.8 to the one of 2007 in Figure 5.10 makes immediately clear how the concept of degree of interest helps us to focus the analysis on relevant data. The DoI-based visualization facilitates a quick overview of those graph elements that are of particular interest to the user. In our case, the top authors are emphasized, while less-relevant data are represented in an aggregated fashion to provide some context.

Here, we discussed the specific example of large dynamic graphs. Yet, the general idea of degree of interest can be applied to other data classes as well. Its most important characteristic is that we can distinguish between relevant and less-relevant data. This allows for emphasizing those data elements that match the interests of the user, while the others are summarized or even omitted.

In the next section, we continue with another approach for narrowing down the analysis on relevant data, more precisely, on features.

5.2.2 Feature-based Visual Analysis

The goal of feature-based visual analysis is to automatically extract features that capture meaningful data characteristics [RPS01]. The visual analysis will then concentrate on the features, rather than on individual data elements. This approach has two key benefits. First, we can reach a higher level of analytic abstraction. And second, the visual representations are clearer and less cluttered, because usually the number of features is much smaller than the number of data elements.

The basic procedure of feature-based visual analysis consists of three steps:

1. Feature specification

2. Feature extraction

3. Feature visualization

The first step is concerned with the specification of criteria that define interesting features. Based on this specification, features are automatically extracted from the data in the second step. This also involves the tracking of features over time and the detection of events in the evolution of features. The third step is to visualize extracted features and detected events. Next, we will explain these fundamental steps in more detail and illustrate them in the context of visual analysis of time-varying reaction-diffusion systems.

Feature Specification

The first step is the specification of meaningful criteria that characterize the features. What meaningful criteria are depends strongly on the data to be analyzed. For certain types of data, there are tailored feature definitions. For example, in flow visualization, where feature-based visual analysis has its roots, features describe critical points, vortices, or shock waves [Pos+03]. However, such a priori definitions do not always exist. For simulations of reaction-diffusion systems, for example, domain experts are interested in features that describe 3D regions with high concentrations of certain particles. However, the

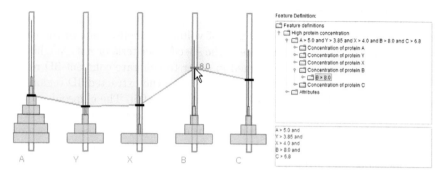

(a) Specification of thresholds.   (b) Formal feature definition.

Figure 5.12   Feature specification with an interactive interface. Courtesy of Christian Eichner.

thresholds of what should be considered a particularly high concentration is not known beforehand, but varies depending on the type of particles and the system being studied. In such cases, features can be specified interactively.

For interactive feature specification, two requirements need to be addressed. First, users must be enabled to graphically input relevant data characteristics as features, and second, the graphical input must map to a suitable formal feature description to enable the later automatic extraction of features [DGH03].

Figure 5.12a shows an interface that supports the interactive specification of features [Eic+14]. The parallel histograms in Figure 5.12a facilitate an informed definition of suitable thresholds. The graphical input performed on the parallel histograms easily translates to a formal definition based on predicate logic. The corresponding formulas describe open or closed intervals of value ranges of interest as combinations of comparison predicates and Boolean logic operators, as illustrated in Figure 5.12b. The formulas can be stored for later reuse and fine tuning. They are also the basis for the automatic feature extraction.

Feature Extraction

Once the feature specification is complete, the next step is to extract instances of features (or simply features) from the data. To this end, the automatic feature extraction is carried out for each type of feature and for each time step in the data. This involves spatial and temporal aspects.

Spatial Aspects In general, the goal is to determine where features are located in the data space and to characterize them by their position, size, shape, and orientation. Again, how the extraction process is implemented depends on the type of features being relevant in the application domain. Dedicated methods exist for the critical points, vortices, and shock waves in flows as mentioned earlier [Pos+03].

For our running example of reaction-diffusion systems, the data consist of particles located in a 3D grid of volumetric cells. Each of the cells is tested whether it matches one of the specified criteria or not. Neighboring cells that do match are merged in order to generate coherent 3D regions representing the features. In a second step, the extracted 3D regions can be abstracted further to 3D ellipsoids [vWal+96]. The ellipsoid axes are oriented according to the eigenvectors of the covariance matrix of the positions of the matching cells, and the length of the axes is determined by the respective eigenvalues. In other words, the ellipsoids appear at locations and stretch out in directions corresponding to the interesting parts of the data.

Temporal Aspects In order to learn about the temporal evolution of features, the ellipsoids need to be tracked over time. The question is which ellipsoid at time t_i corresponds to which ellipsoid at time t_{i-1}? In other words, which ellipsoid at time t_i represents an evolved version of an ellipsoid at t_{i-1}?

By solving this correspondence problem for all time steps, we obtain paths of features over time. Typically, an ellipsoid simply continues to exist with varying position, volume, and orientation, which results in a linear path. Yet, features may also split and merge, or exit and re-enter the data, which might indicate interesting events in the data's evolution.

The detection of such events is an integral part of the feature extraction step. The result of the event detection is a layered graph in which layers are time steps and nodes represent features. Edges exist when there is a correspondence between two features of consecutive time steps. The paths through the graph describe the evolution of features, where particular connectivity patterns represent different events. For instance, a node with one incoming edge and multiple outgoing edges corresponds to a split event.

In summary, the feature extraction produces two results: a set of ellipsoids that characterize features spatially for each time step and an event graph of the temporal evolution of features.

Feature Visualization

The visualization of features is at the heart of feature-based visual analysis. Its goal is to communicate the spatial and the temporal aspects of features as captured in the set of ellipsoids and the event graph.

The spatial aspects can be visualized by showing the 3D ellipsoids of a particular time step as color-coded, three-band contours. The contour rendering helps in reducing 3D occlusion. The color-coding allows us to distinguish different types of features. In our case, different colors represent different proteins. The advantage of this encoding becomes clear in Figure 5.13. In

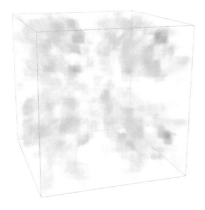

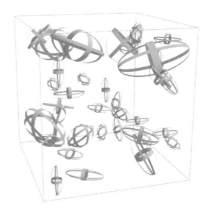

(a) Volume visualization. (b) Ellipsoid visualization.

Figure 5.13 Comparison of direct volume visualization of the particle concentration of one protein and ellipsoid-based visualization of features representing high concentrations of two different proteins. Courtesy of Christian Eichner.

Figure 5.13a, which depicts the concentration of just one protein by direct volume visualization, we can hardly locate and quantify regions with higher protein concentrations. Figure 5.13b visualizes the extracted features instead of the raw data. Red and blue ellipsoids mark regions with high concentration for two different proteins. In comparison to Figure 5.13a, the spatial characteristics of regions with high concentrations can be recognized more easily even for two proteins.

But how do the features evolve? In order to create an overview of the evolution of features across all time steps, the event graph can be visualized as layered node-link diagram. Figure 5.14a shows an example with features of two different proteins. Time is shown from left to right. Each node represents a feature at a particular time point, and the edges show the evolution of features. Symbols indicate interesting events. The size of the nodes corresponds to the extent of the ellipsoids, which makes bigger features easier to spot. Tracing paths of connected nodes allows us to see how features grow, shrink, split, or merge. By comparing sizes and numbers of nodes at a certain time point, it is also possible to estimate whether there are few bigger spots of higher concentration or several smaller ones.

For a detailed comparison, one can visualize the ellipsoids of successive time steps, as shown in Figure 5.14b. It is immediately clear how positions and shapes of features change from one time step to the other. However, this type of visual representation only makes sense when the number of time steps and features is small. Otherwise, the display would be too cluttered to gain any insight.

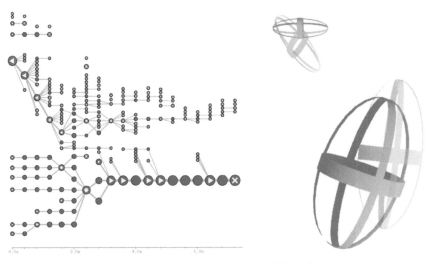

(a) Node-link diagram of the event graph.   (b) Two features at two time steps.

Figure 5.14 Visualizing the temporal evolution of features. Courtesy of Christian Eichner.

To summarize, this section presented the basic steps of feature-based visual analysis. We learned that specifying, extracting, and visualizing features can help us better understand data in their spatio-temporal frame of reference. In the following section, we will discuss the feature-based approach further in the context of a particularly challenging problem: the analysis of chaotic movement.

5.2.3 Analyzing Features of Chaotic Movement

Analyzing chaotic movement is challenging, because interesting patterns are usually covered by loads of irrelevant information. Here, we consider chaotic movement that is generated by a stochastic simulation process. Such simulations are typically controlled by parameters whose influence on the simulation outcome is not clear upfront.

More specifically, the data consist of r simulation runs. Each run $R_i = (P_i, M_i) : 1 \leq i \leq r$ is characterized by a parameter setting P_i (the input of the simulation) and by a corresponding movement M_i (the output). Each movement M_i consists of the trajectories $T_1, ..., T_m$ of m moving entities. The trajectories are sampled uniformly at time steps $t_1, ..., t_n$. Each trajectory point stores various pieces of information, such as position, speed, acceleration, or distance to other entities.

In addition to the spatial and temporal context of the simulated movements, there are also the dependencies on the simulation parameters. There can be thousands of different parameter configurations, each resulting in thousands of

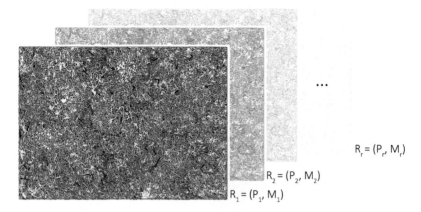

$R_r = (P_r, M_r)$

$R_2 = (P_2, M_2)$

$R_1 = (P_1, M_1)$

Figure 5.15   Drawing thousands of trajectories of chaotic movements for multiple simulations leads to cluttered and indecipherable visual representations. Courtesy of Martin Röhlig.

complex movements with chaotic trajectories. As can be seen in Figure 5.15, indiscriminately drawing all data will certainly lead to extreme visual clutter, which makes it impossible to study individual trajectories or the influence of parameter configurations.

Now let's see how the feature-based approach can help us gain insight into chaotic movement despite its size and complexity. Instead of dealing with individual movement trajectories, we will now work with features of movement.

Feature Specification

The literature on visual analysis of movement data provides various specifications of features that can be interesting to consider [And+13; vLan+14; Lub+15]. Four fundamental categories of features can be distinguished:

- **Basic features** describe aggregated values for all trajectories of the movement such as average speed or average acceleration of the moving entities.

- **Group features** characterize groups of moving entities. Interesting group features are, for example, the number of groups per time step or the ratio of group members and entities not being in any group.

- **Region features** take the spatial distribution of the moving entities into account. For example, it can be interesting to extract regions with high or low density of entities of certain type, and to consider the count, position, and size of these regions.

- **Advanced features** can be derived by further analytical processing of the previous features. As such, they describe features of features.

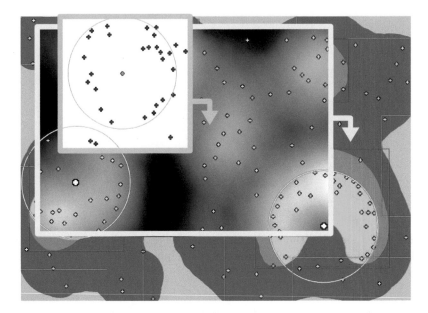

Figure 5.16 From entities (dot marks) to density map (gray-scale image) to regions (colored image). Courtesy of Martin Röhlig.

Advanced features may, for example, detect time points with significant changes of the movement behavior.

A comprehensive visual analysis typically requires several features from all four categories. Which features concretely will be most helpful in a given case depends on the data and the analysis task.

Feature Extraction

The different categories of features listed above require different extraction mechanisms. The extraction of basic features and group features operates on individual trajectories or sets of them. The extraction of region features requires the computation of 2D density maps for each time step of the movement. By quantizing the density maps, regions with different characteristics can be extracted. Figure 5.16 shows a collage for the purpose of illustration. Moving entities and groups are displayed as dot marks and circles, respectively. A density map is visualized as a gray-scale image. Extracted high-density and low-density regions are colored in red and green. Regions that overlap with groups are orange. Blue represents uninteresting parts of the data.

Given the variety of movement features and the complexity of chaotic movement data, it is impossible to carry out the feature extraction on the fly. Instead, an extensive pre-processing step computes as many features as

Figure 5.17   2D movement reduced to 1D time series of feature values.

possible in advance. This enables users to include or exclude certain features on demand during the actual data analysis.

As a result of the feature extraction, the data are reduced considerably. In fact, the complex 2D movements with thousands of entities are condensed down to 1D time series of feature values as indicated in Figure 5.17. These time series are to be represented visually for all simulation runs in the final feature visualization step.

Feature Visualization

The visualization must depict not only the temporal aspects of features, but also the spatial context of the movement and the dependencies on the parameter settings. Because not all information can be squeezed into a single image, an overview+detail design with several linked views makes sense.

Overview   The overview supports the exploration of movement features and parameter dependencies across all simulation runs. To this end, parameter configurations and feature values are shown in a matrix-like fashion as illustrated in the lower left part of Figure 5.18. Parameter values are arranged to the left and the values of a selected feature are shown in the main part of the matrix. A small gap separates parameter values and feature values. The i-th matrix row represents the parameter configuration and the feature time series of the i-th simulation run R_i. The matrix cells are color-coded, where darker colors stand for lower values and brighter colors stand for higher values. Note that the visualization shows only one feature at a time, but interactively switching between different features is not a problem thanks to the feature extraction pre-process.

Detail   The detail views depict temporal and spatial dependencies in more detail, but only for selected parts of the data. As illustrated in the upper part of Figure 5.18, the temporal aspect is communicated by means of a chart view that shows the time series of selected simulation runs. This facilitates comparing the characteristics of the movements under different parametric conditions.

The spatial aspect is detailed by showing selected trajectories for a selected time interval in a trajectory view, which is depicted to the right of Figure 5.18.

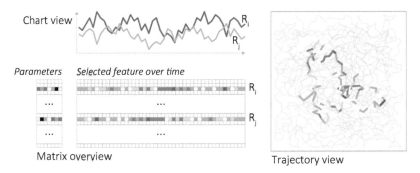

Figure 5.18   Feature visualization with overview and detail views.

The trajectory segments are color-coded, which allows us to see where the movement exhibits certain characteristics. Additionally, the trajectory view can be combined with the computed 2D density map of a selected time step.

Until now, we have sketched the basic ingredients of feature-based visual analysis for chaotic movement data. Implemented and applied to real data, the approach creates visual representations as shown in Figure 5.19. The views are linked, and basic interaction mechanisms allow the user to select particular time series or data ranges to be shown in the detail views. Next, we take a look at some of the insights that could be gained with the feature-based approach.

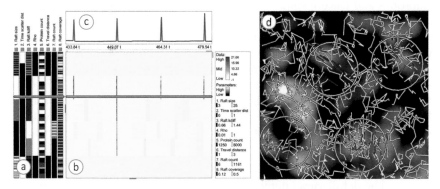

Figure 5.19   Visualization of parameter settings, feature values, and detail information for selected parts of the data. (a) Parameter settings as gray-scale matrix; (b) Feature values over time as color-coded matrix; (c) Chart with selected time series; (d) Trajectory view with selected trajectory segments. Courtesy of Martin Röhlig.

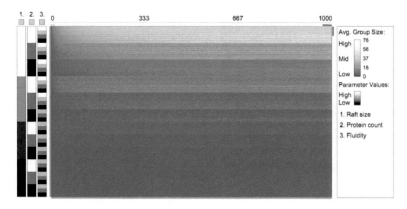

Figure 5.20   Visualizing the parameter dependency of average group size. Courtesy of Martin Röhlig.

Application Example

We already mentioned that the data we are dealing with come from stochastic simulations. The concrete application background is in the context of systems biology where researchers study chaotic movements on cell surfaces. In particular, the researchers seek to gain insight into proteins that move on the cell surface, dock to so-called lipid rafts, move along with them for a short period of time, and then leave them to continue moving freely or dock to other lipid rafts. These dynamic interactions between proteins and lipid rafts play an important role in medicine, for example, in cancer-related research.

In our case, the simulation of the movement is controlled by 8 parameters. Simulations have been carried out for about 2,000 different combinations of parameter values. In each of these simulations, the chaotic movement of about 1,000 lipid rafts and 5,000 proteins has been calculated stochastically over a period of about 4,000 time steps. As a result, the researchers have obtained 180 GB worth of chaotic movement data to be analyzed.

To support different analysis goals for these large data, about 60 features have been computed, including the average speed of all proteins per time step, regions with high and low density of proteins, group characteristics of the lipid rafts, and time periods with relatively constant behavior [Lub+15].

It is beyond the scope of this section to discuss all the features and their visual interpretation in detail. Instead, we want to highlight two findings that could be made by applying the feature-based approach to the described data.

Parameter Dependency Based on the visual representation shown in Figure 5.20, the researchers could detect a dependency of the average group size on the parameters that control the size of the lipid rafts, the number of proteins, and the fluidity of the medium. To make the dependency visible, the rows of the visualization have been sorted with respect to the values of these parameters. The feature-part of the visualization then

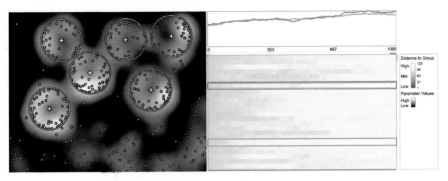

(a) Density map.   (b) Average protein-raft distance.

Figure 5.21   Visualization of the average distance to free proteins reveals the *sweeping effect*. Courtesy of Martin Röhlig.

clearly shows bands with lower group size (darker greens) and higher group size (lighter greens).

Movment Behavior In the visualization shown in Figure 5.21, the researchers could find a behavior called the *sweeping effect*: Regions surrounding the lipid rafts have particularly low density of proteins. The density map in Figure 5.21a makes this obvious for a selected time point.

To confirm that the sweeping effect is permanent throughout the simulation, the feature visualization in Figure 5.21b shows the average distance of free proteins to their nearest lipid rafts over time. As can be seen in the chart view and from the mostly bright colors in the overview matrix, the distance is rather large, and for the two selected runs it even steadily increases slightly toward the end of the simulation.

These two example findings nicely demonstrate that feature-based visual analysis can be very helpful when studying data that are not only large, but also chaotic. The visualization of selected features allows us to concentrate on the important aspects of the data, while leaving aspects that are irrelevant to the task at hand out of consideration.

At this point, we close the section on facilitating the visual analysis by focusing on relevant data. We have learned that four components are necessary in this context. First, we need a way to formally specify what is relevant. For the degree-of-interest approach introduced at the beginning of this section, relevancy is described with respect to data values. The feature-based approach discussed in this section's second part targets higher-level data characteristics.

Second, an automatic computation step has to determine interesting data elements or extract features from the data. This can involve basic comparison operations, but also more complex tracking and event detection methods.

The third component is the visualization of only the relevant parts of the data. Thanks to the reduced amount of information to be communicated, the visual representations are clearer and more focused, which makes gaining insight easier or possible at all.

Fourth and finally, interaction is required to specify one's interest in the data or to select features to be visualized. Interaction adds the flexibility that is necessary to adjust the analysis to changing requirements. In fact, it is again the interplay of computation, visualization, and interaction that makes feature-based approaches so powerful.

There is one aspect that is critical though: In order to focus on relevant data, users must know the target of their tasks. Unfortunately, this is not always the case, particularly when studying unknown data. Therefore, we need further concepts to support the data analysis via automatic computations as we will discuss in the next section.

5.3 ABSTRACTING DATA

We already know that for today's datasets, the number of data values easily exceeds the number of pixels on the display. This might lead to severe over-plotting where very many data values are mapped onto one and the same pixel. Figure 5.22, for example, shows a time series with more than 1.7 million time points. Given the limited width of the chart, each pixel column must accommodate about 1,000 time steps. From a visual data analysis perspective, the question is how to appropriately determine the values that should be visualized in the available display space.

Figure 5.22 Visualization of a time series with more than 1.7 million time points, where each black pixel represents about 1,000 data points. Courtesy of Martin Luboschik.

5.3.1 Sampling and Aggregation

Sampling and aggregation are two approaches that aim at solving this problem. The basic principle of sampling is to select particular data values to be displayed. That is, sampling methods pick from the original data a subset of selected data values. Aggregation methods, on the other hand, combine and reduce several data values to representative values to be displayed. That is, aggregation methods condense the original data down to a reduced set of aggregated values. Note that with sampling, original data values are visualized, whereas with aggregation, derived values representing a couple of original values are visualized.

Sampling In signal processing, sampling aims at representing a continuous signal by a discrete one. To this end, signals are gathered at certain sample points. The Nyquist Shannon sampling theorem tells us that in order to be able to reconstruct the original signal, the number of sample points must be greater than twice the bandwidth of the continuous signal [II91]. In statistics, sampling is understood differently. Here, sampling aims at defining a subset of individuals that are representative of or characterize an entire statistical population [Loh19].

In visualization, both definitions are applied. Sampling refers to gathering data values of variables at certain sample points as well as to gathering data elements. In general, sampling aims at determining a particular data subset that enables the user to see major characteristics of the original dataset. To this end, different sampling strategies can be applied ranging from simple equidistant sampling of data values to sophisticated methods such as graph sampling [LF06]. The used sampling method determines the quality of the output. With simple equidistant sampling, essential data characteristics might be lost. On the other hand, sophisticated sampling methods can preserve essential data characteristics, but they are computationally expensive and time-consuming. Therefore, a balance must be found between the quality of the produced result and the needed computation time.

Aggregation While sampling methods define a subset of data values, aggregation methods replace the data values within an interval by an aggregated value describing certain statistical characteristics of that interval:

- **Minimum:** The smallest value in the interval.

- **Maximum:** The largest value in the interval.

- **Count:** The number of values in the interval.

- **Sum:** The sum of all values in the interval.

- **Average:** The arithmetic mean (the sum divided by the count).

- **Median:** The geometric mean (the central value of the sorted interval).

- **Mode:** The most frequent value in the interval.

- **Count unique:** The number of distinct values in the interval.

- **Standard deviation:** The amount of variation in the interval.

Which aggregation function to use depends on the analysis objective. The demonstrating example in the next section replaces the original data values by maximum and average values.

For very large datasets, a simple data abstraction might not be sufficient. In Figure 5.22, the time series was represented by only one sampling value per 1,000 time points. Such a small subset of samples can hardly represent the whole data faithfully. Therefore, it makes sense to produce multiple data abstractions at different scales. Next, we will discuss such a multi-scale approach in more depth.

5.3.2 Exploring Multi-scale Data Abstractions

Sampling and aggregation can be done repeatedly to produce multi-scale data abstractions. With each repetition, a new coarser scale is created with a decreased number of data points. The creation of multiple scales comes along with two important benefits. First, multi-scale data abstractions allow us to generate scalable and less cluttered visual representations [EF10]. Second, multi-scale data abstractions facilitate the analysis of different characteristics of the data. In fact, coarser data abstractions communicate global trends, whereas less-abstracted scales support the detection of local trends and details. Hence, switching between the scales enables users to gain different insights.

With a larger number of scales, however, it becomes increasingly difficult to inspect all data at all scales. This raises the question of where to drill down from a coarser scale to a finer scale to get to additional information? Because this question is often not easy to answer, it makes sense to guide users to those scales that might yield new findings [Lub+12]. The idea is to exploit data differences between successive scales as an indicator for the availability of additional information at the finer scale. In other words, if two successive scales are quite similar, studying the finer scale might not reveal much new. In contrast, if the two scales exhibit larger differences, it makes sense to drill down and search for local patterns that only take shape at the finer scale.

As an example, Figure 5.23 shows two line charts of a time series at two successive scales. The finer scale at the bottom consists of eleven sample points, whereas the coarser scale at the top has only six points. In the left part, both charts differ only marginally: Both scales show monotonically increasing values. Thus, we can conclude that for the first part of the data, the finer scale does not add substantially new behavior and is therefore less worth exploring. However, for the right part of the charts, things are different. The finer scale depicts a considerable local minimum that is not present at the coarser scale. Identifying and emphasizing such deviating data behavior can in fact guide the user in exploring multi-scale data.

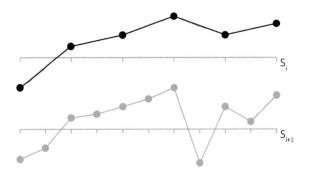

Figure 5.23   One and the same time series at two different scales.

Guiding the User

The simple example of Figure 5.23 illustrates that differences between scales might be a good indicator where further insights can be found. Next, we discuss how such differences can be extracted and visualized along with the abstracted data.

Extracting Differences Between Scales   The extraction of differences is done for all pairs of successive data scales. For each pair, the following three steps are carried out: map sample points, compute difference measures, and aggregate differences.

1. **Map sample points.** First, the sample points at the two successive scales must be unified. To this end, sample points that exist only at one scale are mapped to the respective other scale. This typically involves the computation of new data values, for example, by linear interpolation. The example in Figure 5.24 illustrates how every second sample point of the finer scale is mapped to a new sample point with an interpolated data value at the coarser scale. In the end, both scales have the same sample points, but different data values.

2. **Compute difference measures.** Computing the differences between the two scales is the second step. Various measures can be used to quantify differences. Figure 5.25 illustrates two examples, a point-wise measure and a segment-wise one. A simple point-wise measure is to compute the *absolute value difference* (AVD) per sample point, for example, by means of the Euclidean distance. A simple segment-wise measure is the *slope sign difference* (SSD). It captures whether the signs of the slopes of two segments differ. In Figure 5.25, the signs are indicated per segment as − and +. If the signs differ, the slope sign difference is 1, otherwise it is 0. While being computationally inexpensive, these simple measures capture just enough information to be useful as indicators.

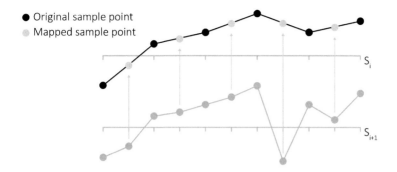

Figure 5.24   Unifying the sample points of two successive scales by mapping and interpolation.

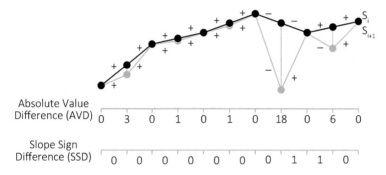

Figure 5.25   Computing the absolute value difference (AVD) and the slope sign difference (SSD) between two successive data scales.

3. **Aggregate differences.** Because our goal is to guide users to regions worth studying in detail, we need to aggregate the point-wise or segment-wise difference values over larger intervals. Figure 5.26 shows a maximum aggregation of AVD and an average aggregation of SSD as examples. A maximum aggregation ensures that larger differences are not compensated by smaller ones within the same interval. If only the existence or absence of differences between scales is captured, the average can indicate how much additional information will appear when zooming into a certain range. In our example, both aggregations suggest drilling down primarily into the interval where the local minimum would be revealed.

The outlined basic approach is freely adaptable. Different methods can be applied for mapping the sample points and deriving the missing data values, for measuring the differences between scales, and for aggregating the differences. What remains to be done is to visualize the aggregated differences along with the actual data.

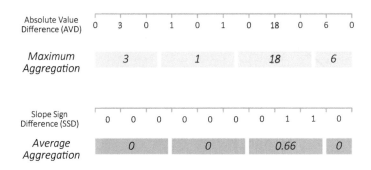

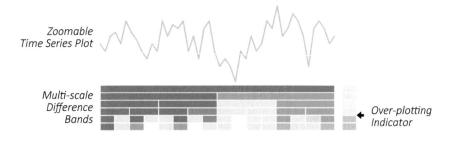

Figure 5.26 Aggregation of data differences with maximum aggregation for the absolute value difference (AVD) and average aggregation for the slope sign difference (SSD) function.

Visualizing Aggregated Scale Differences To guide users in exploring multi-scale data, the aggregated differences are visualized along with the data. A schematic illustration of the approach is given in Figure 5.27. The main time series plot is accompanied by a stack of so-called difference-bands. Each band is further sub-divided into color-coded cells representing the aggregated differences between the adjacent data scales. There are two modes for the color-coding: global and local. For global color-coding, colors are mapped between the global minimum and maximum value of all cells. This facilitates comparison across scales. For local color-coding, colors are mapped with respect to the minimum and maximum value per band. This makes it easier to see smaller differences per individual scale. In both modes, brighter cells indicate regions with no or only small differences between scales. Darker cells represent regions where the finer scale is substantially different from the coarser scale, and hence, might be interesting to inspect in more detail.

Additionally, an over-plotting indicator shows, depending on the current display resolution, at which scale over-plotting will occur. It is reasonable to start the data analysis with the finest scale that is still free from over-plotting.

Figure 5.27 Visualizing aggregated differences along with the actual data.

From there, the user can zoom and pan the time series plot to reach different scales and different regions as suggested by the color-coded cells. Next, we will illustrate how the described approach can support the visual analysis of a large time series from systems biology.

A Use Case from Systems Biology

Our example is based on time-series data that result from a simulation of the cell division cycle in fission yeast [Lub+12]. For our purpose, it is sufficient to know that the simulated model consists of two different proteins that control the major events of the cell division. The simulation produces 3.6 million data points that describe the dynamic change of molecule quantities over time. By repeated average aggregation, the original data are transformed into a multi-scale data abstraction with 21 levels. Differences between the scales are computed by the measures introduced before: slope sign difference (SSD) with average aggregation and absolute value difference (AVD) with maximum aggregation.

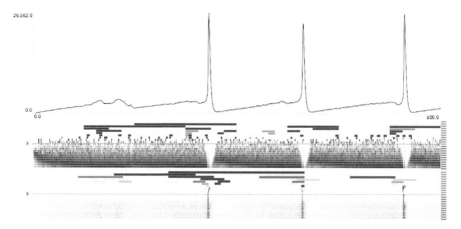

Figure 5.28   Time series plot of the simulation outcome and corresponding multi-scale difference bands of SSD and AVD with local color mode. Courtesy of Martin Luboschik.

The outcome of the simulation is visualized in Figure 5.28. The time series shows considerable oscillations. To get more insights, we check the multi-scale difference bands of SSD (top in blue-yellow) and AVD (bottom in green-yellow). Particularly, at the finer scales that would produce over-plotting, as signaled in red color in the over-plotting indicator, we can observe something interesting. For the SSD measure, we can see for each peak in the time series a corresponding brighter notch in the otherwise rather bluish difference bands. Within each notch, there seems to be a very thin blue upward spike, which would suggest that some additional information can be found. This is confirmed

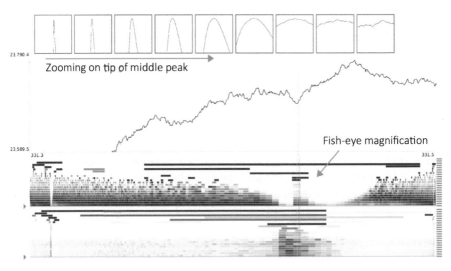

Figure 5.29 Studying the details of the middle peak-notch-spike pattern from Figure 5.28. Courtesy of Martin Luboschik.

by the AVD measure, where green downward spikes are clearly visible. This prompts us to look into these parts of the data at a finer scale. We decide to study the middle peak-notch-spike pattern in more detail.

In Figure 5.29, we perform two separate drill-down operations. First, we enlarge the multi-scale difference bands where the middle notch is located by applying a horizontal fish-eye magnification (a 1D variant of the fish-eye lens introduced in Section 4.6.3 of Chapter 4). We can see, yes, there are indeed some significant differences in this region as indicated by the darker, more saturated colors. Second, we switch the time series plot to the finest data scale and zoom far in on the tip of the peak. This reveals the source of the differences: There are considerable fluctuations at the tip. These are caused by the two proteins competing with each other in the simulated process. We can conclude that there is no clear tipping point as the coarser scales might have suggested. This is an interesting finding that would likely have gone unnoticed without the guidance offered by the multi-scale difference bands.

In summary, data abstraction by means of sampling and aggregation is an important approach to dealing with very large data. However, representing millions of data points by a comparatively small number of samples or aggregated values does not adequately communicate the complete data behavior. Therefore, multi-scale data abstractions are required. They facilitate the analysis of data behavior at different scales and help us generate scalable visual representations. Computing and visualizing differences between subsequent scales helps users identify those scales and ranges of the data that should be investigated in detail.

While sampling and aggregation are not very selective with respect to which data elements are abstracted, we will next look into methods that particularly focus on data elements that are similar.

5.4 GROUPING SIMILAR DATA ELEMENTS

With sampling and aggregation, data values within a certain data interval are replaced by a single representative value. Now, we consider the grouping of similar data elements. Classification and clustering are typical approaches for grouping data elements. While the goal of grouping is the same for both approaches, they achieve the goal in different ways. Classification subdivides the data space, whereas clustering partitions the set of data elements. A class describes a subspace of the data space, whereas a cluster represents a subset of similar data elements. Classes are characterized by value ranges, whereas clusters are characterized by the properties of their affiliated data elements. Representing a dataset by classes or clusters significantly reduces the data complexity and facilitates the generation of overviews. In the following, we will discuss classification and data clustering in more detail.

5.4.1 Classification

Classification subdivides the data space. Each subspace defines a particular class. Data elements falling into the same subspace share the same data characteristics and are said to be in the same class. The key to classification is a suitable description of the classes.

Basic Means of Specifying Classes

There are different strategies to define the classes. A very basic approach is to divide the domain of the data variables into intervals. For example, the age of persons in a social network could be grouped into four classes: kids ($age \leq 12$), youths ($12 < age \leq 18$), adults ($18 < age \leq 60$), and seniors ($60 < age$). This reduces the range of age values, usually between 0 and 110, to only four class values. Another example is to group the wind direction into eight major directions: N, NE, E, SE, S, SW, W, and NW. Here, the potentially continuous angle value between $0°$ and $360°$ is reduced to eight class values.

A common approach to classifying multivariate data elements is to use *decision trees*. Decision trees partition the data space based on a series of tests. For quantitative or ordinal data variables, tests are usually comparisons against given thresholds. Categorical data can be partitioned by means of yes/no questions. The specified tests are the basic building blocks of decision trees. A test defines a node in a decision tree. For each class distinguished by a test, there is an edge attached to the test's node. The edges connect a test to follow-up nodes in the decision tree. One test is chosen to be the root node of the decision tree. The leaves of the decision tree are the produced classes. A

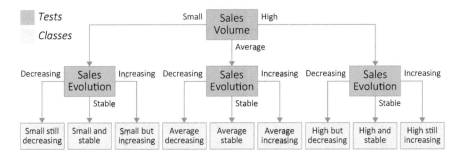

Figure 5.30 A decision tree classifying enterprises according to their sales.

data element is assigned to a class, if the data element matches all the tests along the path from the root node of the decision tree to the class leaf node.

Visualizing decision trees can help users understand how the test criteria connect to a hierarchy of decisions that result in the partitioning of the data space. For this purpose, common tree visualization methods can be applied, for example, node-link diagrams as introduced in Section 3.5 of Chapter 3. A simple decision tree for a dataset with enterprise sales volumes is depicted in Figure 5.30. The first test in the root node refers to sales volumes. It distinguishes enterprises with small, average, and high sales. The next hierarchy level considers the evolution of sales to further distinguish enterprises with increasing, decreasing, or stable sales. In the end, we obtain a grouping of the enterprises into nine classes: small still decreasing, small and stable, small but increasing, and so forth. Each class represents enterprises that share the same characteristics.

Note that in other scenarios, decision trees are not necessarily leveled and balanced as in our example. Moreover, the structure of how tests lead to a decision for one class or the other can be significantly more complex. In fact, the specification of adequate test criteria for complex domain problems can be quite challenging. There are even scenarios where it is not possible at all to formulate the classification as a decision tree. In such cases, sophisticated algorithms and careful parameter tuning are required in order to arrive at appropriate classification results. Next, we will discuss such a sophisticated classification approach and illustrate how visual analysis can help to tune and evaluate the classification outcome.

Visual Classification Support: An Example from Activity Recognition

The recognition of activities is a relevant task in various domains. For example, assistance systems observe people's activities in order to automatically offer on-demand help [Tei+17]. To this end, people are instrumented with several sensors that produce large volumes of time-series data. The task of classification in this context is to deduce from the multivariate sensor data the activity that

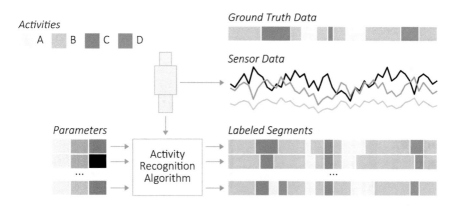

Figure 5.31 Illustration of activity recognition based on parameter-dependent algorithms that learn from some ground truth. Adapted from [Röh+15].

a person is or was doing. A very simple example would be to label certain intervals in time as walking activities. Yet, an activity can also be to cook a meal, which is certainly not so easy to discriminate.

In general, recognizing activities requires sophisticated data classifiers. A typical way how such classifiers work is to learn from human-labeled ground truth data. The ground truth is collected in observational studies where a person's activities are video recorded in addition to being monitored by the sensors. Based on the video, a human can easily label each characteristic sequences of the observation as a certain activity, for example, sitting, standing, or walking. The assumption is that the sensor data associated with the manually labeled activities are discriminatory enough so that a classifier can learn from the sensor data how to recognize activities robustly.

Yet, as mentioned before, the algorithms involved in activity recognition typically depend on careful parameter tuning. To obtain accurate results, the algorithms need to be tested with different parameter configurations. The goal is to find configurations that produce classifications that match the ground truth best. Figure 5.31 illustrates the described situation. Next, we explain in more detail how visual analysis methods can help users understand the influence of parameters on the classification results. The better understanding makes it easier to adjust and fine-tune the parameters to achieve a better classification quality.

Following the previous schematic illustration, Figure 5.32 shows a concrete visualization of the classified sensor data in concert with the underlying parameter settings and the ground truth [Röh+15]. The visual design is similar to the matrix-based visualization for feature visualization as described back in Section 5.2.3. The left part of the visualization (a) shows the parameter configurations. Each column corresponds to a parameter, and each row represents the values of a parameter configuration. The main part of the visualization (b) shows the activity-labeled time series as colored pixel rows from left to

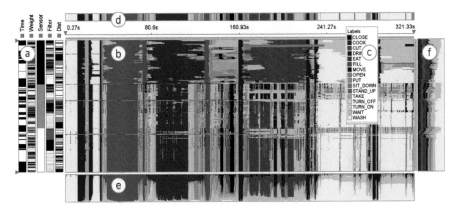

Figure 5.32 Visualization of parameter-dependent classification outcomes for activity recognition. (a) Parameter configurations; (b) Recognized activities; (c) Color legend; (d) Ground truth; (e) and (f) Stacked histograms with aggregated information. Courtesy of Martin Röhlig.

right. For each parameter configuration, there is a corresponding pixel row in the main part, and each activity is painted with a distinct color (c). This visualization provides a nice overview of the classification outcome for all different parameter configurations. Moreover, we can easily see which activity has been performed at which time step.

Figure 5.32 also shows the ground truth as a separate colored band at the top (d). This facilitates the visual comparison of the recognized activities and the actually performed activities. Additionally, stacked histograms (e) and (f) are attached at the bottom and to the right. They show the distribution of detected activities per time step and per parameter configuration, respectively. This gives us a sense of the stability of the activity recognition, both over time (e) and over the different parameter configurations (f).

For evaluating the activity recognition in more detail, it makes sense to visually analyze the discrepancy between the detected activities and the ground truth. To this end, the color coding and the ordering of rows is changed in Figure 5.33. Instead of showing distinct colors per data class, the visual representation now shows red color where the recognized activity does not correspond to the ground truth. We can now see that the number of incorrect classifications increases over time for the majority of parameter setting and that the overall quality of the activity recognition depends mostly on the parameter sensor.

While visual analysis cannot improve the classification outcome, it nonetheless helps users to understand why the activity recognition behaves the way it does. The insights generated with the visualization can be utilized to estimate the influence of parameters and to fine-tune parameter configurations accordingly. Parameters with little impact can be paid less attention. For parameters

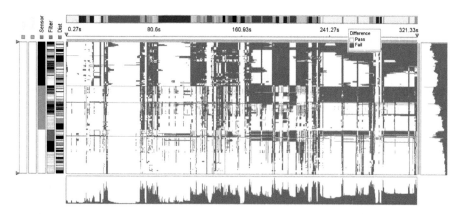

Figure 5.33 Highlighting the incorrectly classified time steps in red. Courtesy of Martin Röhlig.

with substantial impact, the visualization can indicate parameter ranges that are good candidates for further refinement in search of better classification results.

In summary, this section taught us that classification is about grouping data by dividing the data space into subspaces with specific data characteristics. Classification can be as easy as testing specified criteria, possibly organized in a hierarchical schema. Yet, classification can also be so complex that sophisticated algorithms need to be involved. In both cases, visualization, either of the classification schema as for the decision trees or of the parameter-dependent classification outcome as for the activity recognition example, can help users understand how the data are grouped into classes. Next, we study clustering as another common approach to group similar data elements.

5.4.2 Data Clustering

The goal of clustering is to group similar data elements into clusters. Data elements in the same cluster are similar with respect to a specified similarity measure. The elements of different clusters should be dissimilar. Clustering can support the visual analysis in different ways. Data elements can be arranged or colored according to their cluster affiliation. Overview representations can concentrate on the clusters, rather than on individual data elements. Clustering can also help us study the data at different levels of abstraction. All this can reduce the analysis complexity significantly.

Configuring the Clustering

Data clustering can be done in various ways. Therefore, it is first of all necessary to configure the clustering, that is, to specify what to cluster and how to cluster.

Let us next briefly discuss the decisions that need to be made before the actual clustering can be performed.

What to cluster? First, we must decide *which data elements* should be taken into account. Redundant data elements and data elements that represent outliers can distort the grouping, and thus they are usually excluded from clustering.

Second, in the case of multivariate data, we must decide *which data variables* should be taken into account. Should we consider all variables or only selected ones? We must also decide whether the variables should be weighted equally or whether some of them should be given higher or lower weight. Variables that do not contribute to the main topic of interest could be ignored or assigned a lower weight. Highly relevant variables should get higher weights to give them more influence on the clustering outcome.

For example, consider a dataset that contains information about companies, their name and field of business, chief executive officer, sales, revenue, balance sheet total, and number of employees. If the analysis objective is to cluster highly profitable and big companies, the attribute revenue should be prioritized, the attributes sales, balance sheet total, and number of employees should be included, and the attributes name, field of business, and chief executive officer can be neglected.

How to cluster? Third, we have to decide on the *similarity measure* to be used to determine whether data elements are similar. A widely applied measure for quantitative data is the Euclidean distance, which describes the straight-line distance between two data elements. The Manhattan distance defines the distance between two data elements as the sum of the absolute difference of their data values. Further measures for quantitative data are the cosine distance or correlation coefficients such as the Pearson correlation. For qualitative data, other distance measures must be applied, for example the Hamming distance, the Levenshtein distance, or the Jaccard index.

Continuing our previous example, for clustering companies according to profits and size, the Euclidean distance would be an appropriate measure. Yet, things are different when we change our analysis objective. Let's now assume that the business data are given on a yearly basis. If our analysis objective is to study companies with similar development over time, a correlation-based measure would be appropriate. This way, the clustering would group companies that have similar development histories.

Finally, we must decide on the *clustering strategy* to apply. The clustering strategy defines how the data elements being similar according to the chosen similarity measure are grouped into clusters. There are four basic clustering strategies: partitioning clustering, hierarchical clustering, grid-based clustering, and density-based clustering [HKP11]. Each clustering strategy takes a different approach and comes with its own set of parameters, as we will see in a moment.

Taken together, excluding data elements from clustering, weighting variables, using different similarity measures, applying different clustering strategies, and choosing different parameters, all these decisions must be carefully weighed. It is important to realize that slightly different decisions can produce totally different clustering results. Which configuration to employ depends on the data and the objective of the analysis.

Computing the Clusters

In the following, we will explain how clusters of similar data elements can actually be computed. To this end, we will briefly describe the four different clustering strategies just mentioned.

Partitioning clustering algorithms generate $k < n$ groups from n data elements by partitioning the data space. The most popular method is the *k-means* algorithm. It describes each cluster by its centroid. Initially, the k centroids are distributed randomly in the data space. Then, the data elements are assigned to their closest centroid. The positions of the centroids are then updated to the average position of the data elements within a cluster. After updating the centroid positions, it can happen that for some data elements, their assigned cluster centroid is no longer the closest one. This requires a reassignment of the data elements to the cluster that is now the closest. These changes again imply re-calculations of the centroid positions. This iterative clustering process stops when the centroid positions have settled at sufficiently stable positions.

The k-means algorithm is popular for its simplicity and general applicability. However, a difficulty is that the number of clusters must be specified in advance, which can be problematic for unknown data.

Hierarchical clustering algorithms organize nested clusters in a hierarchy. The similarity of data elements within the clusters increases as the hierarchy is descended. The root node of the hierarchy represents the cluster that contains all data elements. The leaves of the hierarchy represent clusters that contain only a single data element.

A horizontal cut through the hierarchy describes a particular level of abstraction of the clustered data. Defining an appropriate cut is the difficulty of hierarchical clustering. Here, visualizing the cluster hierarchy as a so-called *dendrogram* can help users find a good level of abstraction. Cutting the hierarchy near the root produces an overview of the data. Cutting closer to the leaves delivers more details.

Hierarchical clustering can be implemented in two ways. *Agglomerative algorithms* recursively group the two most similar data elements in a bottom-up fashion. A popular example is Ward's method. In contrast, *divisive algorithms* recursively split the data into two dissimilar subsets in a top-down fashion until each data element forms its own cluster.

Grid-based clustering algorithms subdivide the data domain by means of a regular grid. Each data element falls into a particular cell of the grid, and for each grid cell statistical meta-data, such as minimum, maximum, or average, are computed. Based on a data query, clusters are formed by connecting those grid cells that contain a sufficient number of data elements that match the query. Obviously, the grid resolution is a critical parameter in terms of the clustering quality. STING (statistical information grid) is an example that uses a hierarchical structure of grid cells. Such multi-resolution grids make it possible to formulate cluster queries against very large databases.

Density-based clustering algorithms distinguish between regions with higher and lower density, and define clusters based on this distinction. DBSCAN (density-based spatial clustering of applications with noise) is a prominent example of density-based clustering. DBSCAN groups data elements based on two criteria. The data elements must be close to each other (distance) and there must be sufficiently many data elements around the same spot (density). Data elements that lie in sparse regions are considered noise or outliers. Being robust against outliers is a particular advantage of this method. On the other hand, DBSCAN requires careful tuning of the distance and density criteria in order to be able to detect clusters. It works less well with evenly distributed data.

Figure 5.34 illustrates the four clustering strategies applied to an artificial bivariate dataset. The data (a) contain two well-separated subsets and also two major outliers and two minor outliers. It is interesting to see how the different clustering algorithms cope with these specific data characteristics.

For the k-means clustering, setting $k = 2$ seems to be appropriate, since we can see that there are two distinguishable subsets. Similarly, cutting the dendrogram (f) near the root to obtain two clusters for Ward's method seems to make sense. However, the two clusters extracted in (b) and (c) do not correspond to the two obvious subsets in the data. Being a partitioning approach, k-means generates convex clusters. Moreover, the major outliers distort the clustering of k-means and Ward's method.

In contrast, STING (d) and DBSCAN (e) produce two clusters representing the two obvious subsets quite well. Yet, STING also generates clusters for each of the four outliers. DBSCAN treats the outliers differently. The major outliers are recognized as such. The minor outliers, however, are assigned to the two clusters. In order to keep them as outliers too, the distance parameter of DBSCAN could be decreased. However, only carefully so, because the clusters could fall apart if the distance parameter is too low.

In light of these examples, it is clear that there is no silver bullet strategy to clustering. Instead, it is necessary to configure the clustering appropriately to obtain suitable results. However, it is often unclear how the clustering result should look like, especially when analyzing unknown data. Similarly, how different configurations affect the result is often unclear as well. Again,

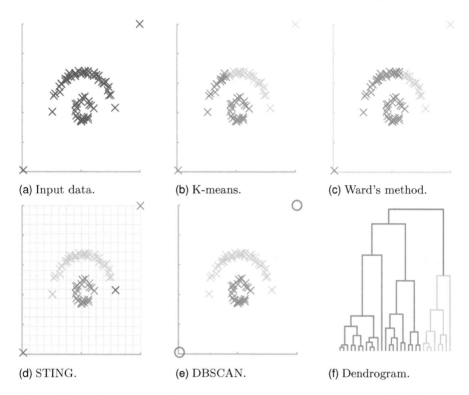

(a) Input data.

(b) K-means.

(c) Ward's method.

(d) STING.

(e) DBSCAN.

(f) Dendrogram.

Figure 5.34   Illustration of clustering strategies. Adapted from [Gla14] with permission of Sylvia Saalfeld.

interactive visual methods can help us perform the clustering appropriately, check the computed clusters, and, if necessary, to specify a new configuration that delivers more plausible results.

StratomeX is a tool that supports the direct visual comparison of clustering results generated with different algorithms and configurations [Lex+12]. Figure 5.35 shows the interface of StratomeX. The columns represent the same microRNA data, but differently clustered by three algorithms. Bands between the columns connect the same data elements. We can nicely see that data elements being together in a group for one clustering are distributed across several groups for another clustering. For example, the data elements of the highlighted first group of the k-means clustering (middle column) disperse into three different groups for the hierarchical clustering (left column). On the other hand, a large portion of the k-means group also forms a large portion of an affinity propagation cluster. Based on these observations, the user can inspect the data elements in the marked clusters in more detail in order to judge whether the produced clusters are in fact adequate groupings.

Finally, it is worth mentioning that in addition to the illustrated fundamental clustering strategies, there are further ways of clustering data. For example,

(a) Hierarchical clustering.   (b) K-means clustering.   (c) Affinity propagation.

Figure 5.35 Comparison of clusters generated with hierarchical clustering, k-means, and affinity propagation. Software courtesy of Alexander Lex.

distribution-based clustering groups data elements according to statistical distribution models. Spectral clustering works on the spectrum of eigenvalues of the similarity matrix of the data. Clustering can also be done with neural models, such as self-organizing maps, as we will see next.

Hybrid SOM-based Clustering

Self-organizing maps (SOM) are a form of artificial neural networks. SOMs can be applied to all kinds of data analysis problems. Here, we utilize them for grouping similar data elements. SOMs are suited for finding groups even in large and unstructured data.

Basic SOM Clustering   A SOM basically corresponds to a regular grid of neurons, each being represented as a so-called *reference vector*. The number of components per reference vector, that is, the dimensionality, corresponds to the number of data values per data element.

SOMs operate in two steps: training and mapping. In the training phase, the grid of neurons is trained based on so-called *input vectors*, which in our case is a subset of the data elements to be clustered. In the mapping phase, the remaining data elements are processed.

In the training phase, the following four steps are performed to align the reference vectors with the input vectors (the training data):

1. **Initialization** The reference vectors of all neurons in the grid are initialized with random values.

2. **Competition** For each input vector, all neurons in the grid compete for the assignment. The neuron with the most similar reference vector is the winner neuron. The corresponding reference vector is updated such that it more closely represents the input vector.

3. **Cooperation** The winner neuron stimulates neurons in its local neighborhood. The stimulus impact decreases with the training duration.

4. **Adaptation** The reference vectors associated with the stimulated neurons are adjusted with regard to the input vector. The degree of adaptation also decreases with the training duration.

After the training phase, the grid's reference vectors reflect a similarity-based arrangement of the input vectors. Adjacent grid cells with similar reference vectors represent groups of similar data elements. Then, in the mapping phase, the remaining data elements, that is, those data elements that were not used in the training phase, can be assigned to the closest matching neuron.

With the outlined procedure, SOMs can produce useful groupings of data elements. Yet, it is not easy to determine the appropriate number of neurons (or reference vectors) required to generate results that are balanced in terms of clustering quality and computation time. The proper number of neurons is not known in advance, as it depends heavily on the given data. With too few neurons, the clustering quality will be low. With too many neurons, the quality might no longer improve, but the computation time will be higher, which might impair the interactive visual data analysis.

Hybrid Clustering with SOM   To keep computational costs and cluster quality balanced, it makes sense to implement a hybrid clustering procedure that combines SOM clustering with a second clustering algorithm [JTS08]. Such a hybrid clustering involves two steps. First, a coarse SOM training generates a moderate number of SOM clusters. Then, a second clustering algorithm, such as hierarchical clustering, further refines each SOM cluster.

Figure 5.36 illustrates how such a hybrid clustering can be employed to sort the rows of a *table lens* visualization, a tabular data representation that we introduced in Section 3.2.1 of Chapter 3. The goal is to order the rows of the table such that similar data tuples are arranged closely together. To this end, a SOM delivers a rough grouping of rows in the first step. In the second step, hierarchical clustering creates a cluster hierarchy per SOM cluster. Figure 5.36b shows the ordered table lens after the hybrid clustering has been applied. While the original arrangement of rows in Figure 5.36a exhibits many color switches, the hybrid clustering leads to fewer color switches across rows. Additionally, the right-most column in the table in Figure 5.36b shows the cluster hierarchies as *icicle plots*. This visual representation can be utilized to adjust the visualized level of abstraction by expanding and collapse clusters.

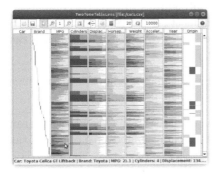 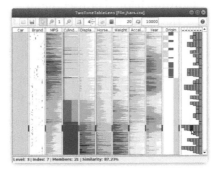

(a) Unordered rows before clustering.   (b) Ordered rows after clustering.

Figure 5.36 Applying hybrid SOM-based clustering to sort rows in a table lens visualization.

Thanks to the hybrid nature of the procedure, the clustering is computationally efficient. Since we need only a small number of neurons for the initial rough SOM grouping, computation time is reduced. The same applies to the hierarchical clustering part. Because we do not compute a global hierarchy, but refine each SOM cluster separately, the individual runtime is reduced. Overall, combining clustering methods to a hybrid approach can help us balance clustering quality and computational costs.

The clustering strategies described so far are generally applicable to multivariate data. Yet, when we go beyond multivariate data, it is often necessary to employ advanced clustering strategies that are dedicated to the peculiarities of a specific class of data. The next section will introduce such an advanced strategy for the particular case of clustering multivariate dynamic graphs.

5.4.3 Clustering Multivariate Dynamic Graphs

Clustering graphs is a complex problem [Sch07]. The goal is to group the nodes of a graph into clusters based on some notion of similarity. The clusters can then be represented by so-called meta-nodes, which basically subsume all nodes belonging to a cluster. Visualizing the meta-nodes facilitates an overview on key characteristics of a graph. By applying the clustering repeatedly on nodes and meta-nodes, a clustering hierarchy can be generated to facilitate multi-scale data analyses, which is particularly useful for large and complex graph structures.

In the following, we will study ways to cluster multivariate dynamic graphs, that is, graphs whose nodes have data attributes and also references to time. A multivariate dynamic graph DG can be denoted as a sequence of graphs $DG = (G_1, G_2, \ldots, G_n)$, where $G_i = (V_i, E_i)$ is a graph with multivariate nodes V_i and edges E_i at time $t_i \in T$.

There are two basic clustering strategies taking into account the time-varying character of multivariate dynamic graphs [Had14]. *Attribute-driven methods* group nodes according to the temporal evolution of their attribute values. In contrast, *structure-driven methods* cluster a graph based on its structure as defined by the time-depended existence of nodes and edges.

Attribute-based Clustering

Attribute-based clustering operates on the super-graph $SG = \bigcup G_i$, which subsumes all graphs $G_i \in DG$. The goal is to extract from SG clusters whose attributes share similar temporal behavior, for example, stable, increasing, decreasing, or recurring values. This can be achieved in three steps: preprocess, configure, and cluster.

1. **Preprocess** First, the attribute values associated with the graph elements must be transformed into time series that describe the temporal behavior. To this end, the attribute values are concatenated according to the sequence as defined by DG. However, the resulting time series might contain missing values, because nodes and edges do not necessarily exist at all points in time. These missing values are unambiguously marked with a dedicated null value ω.

   Additionally, minor fluctuations in the time series should be filtered out in order to reduce the influence of noise on the clustering results. In the end, we obtain for each graph element and for each attribute a complete and smoothed time series with exactly one value (actual data or ω) per time point.

2. **Configure** The second step is to configure the clustering, which in the first place means deciding which attributes should be taken into account. Moreover, the role of the ω values must decided. Including or excluding ω leads to different clustering outcomes. If the ω values are included, they dominate the clustering of those nodes that exist only for a few time points. Such rare nodes will then appear in a distinct cluster, which can be intended. On the other hand, excluding the ω values leads to clusters that contain nodes with similar temporal behavior, which is what we typically want.

   The configuration also involves a decision on the similarity measure and the clustering method to be used. Typically hierarchical clustering is applied. However, other methods can be used as well, in particular methods that are tailored to clustering time series [Lia05].

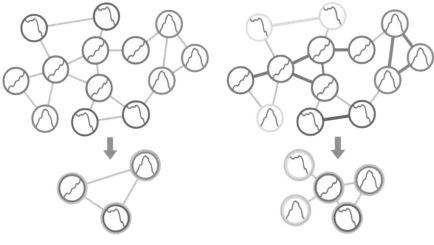

(a) Initial grouping.   (b) Refined clusters.

Figure 5.37   Two-step procedure of clustering nodes based on their attributes. First, nodes with similar attribute behavior are grouped. Second, groups are refined based on connected components. Adapted from [Had14].

3. **Cluster** Finally, the super-graph SG is actually clustered using a two-step procedure. It is illustrated in Figure 5.37 with a node-link visualization whose 13 nodes incorporate miniature charts of their attribute values over time. The first step is to group nodes with similar time series as in Figure 5.37a. Our example shows three preliminary groups (in red, green, and blue) and the resulting abstracted graph at the bottom with three meta-nodes. However, the green and blue groups subsume nodes that are actually not directly connected in the super-graph SG. This is not yet the desired result.

Therefore, the structure-preserving second step refines the groups with respect to their intra-group connected components. Figure 5.37b visualizes the resulting clusters by varying edge colors. The red group has only one connected component. Yet, the green group and the blue group each consist of two connected components, indicated by brighter and darker shades of green and blue, respectively. As the final result, we obtain an abstracted graph that better captures the original connectivity and is still considerably reduced to five meta-nodes.

The extracted attribute-based clusters can now be used to visualize an original multivariate dynamic graph in an abstracted fashion. To this end, the visualization displays the meta-nodes and the corresponding meta-edges and only traces of the original data. Figure 5.38 shows such an abstracted representation. The meta-nodes are represented as small boxes depicting the time series that is representative of the corresponding cluster. A label shows the

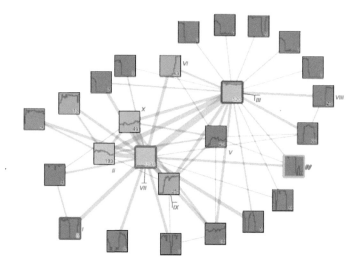

Figure 5.38   Visualization of an attribute-based clustered multivariate dynamic graph. Reprinted from [Had+13].

number of original nodes subsumed in a meta-node. The box colors encode the intra-cluster similarity of the underlying original time series, with darker colors symbolizing larger similarity. The meta-edges visualize connections between the clusters. The edge width encodes the number of original edges subsumed in a meta-edge. Overall, this abstracted visual representation can help us gain an overview of the key temporal behaviors and structural relationships of multivariate dynamic graphs.

Structure-based Clustering of Dynamic Graphs

Structure-based clustering aims to extract key characteristics of the structural changes that occur in a dynamic graph over time. In contrast to the previous section, we now group graphs $G_i \in DG$ with similar structure, rather than nodes with similar attribute behavior. In other words, our clusters now contain a number of similar graphs. As such, a cluster can be understood as a particular *state* that the dynamic graph assumes at certain periods in time. Consequently, the changes in the structure of a dynamic graph can be understood as *transitions* between states. Following this thinking, the idea of structure-based clustering is to construct a state-transition graph that can help us understand how the structural aspect of a dynamic graph varies temporally.

Computing the state-transition graph involves two steps: the initial setup of states and transitions, and the subsequent hierarchical grouping of similar states to meta-states and the corresponding definition of meta-transitions. The overall procedure is illustrated in Figure 5.39 and will be explained in more detail next.

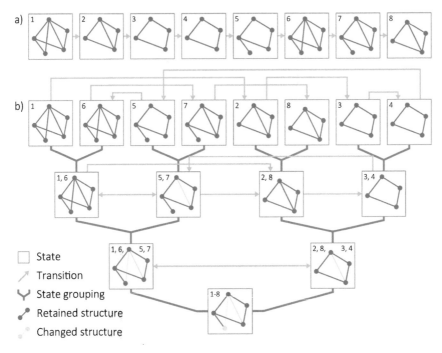

Figure 5.39 Structure-based clustering. (a) Initial set of states and transitions based on the sequence of graphs $G_i \in DG$; (b) Hierarchical grouping of states and transitions based on similar structures. Adapted from [Had14].

1. **Initial setup.** Initially, each original graph $G_i \in DG$ is represented by a separate state. Any two states accommodating successive graphs G_i and G_{i+1} are connected by transitions. The created states and transitions as shown in Figure 5.39a are the basis for the subsequent grouping.

2. **Hierarchical grouping.** For the actual grouping of states, their pairwise similarity needs to be calculated. This can be done in different ways, for example, by means of the *graph edit distance*, which basically counts the number of inserted, deleted, and substituted nodes and edges.

   States with high similarity can be subsumed to form new meta-states. Figure 5.39b illustrates this with pairs of states. The new meta-states themselves can in turn be grouped into further new meta-states. In our example, this is done in a level-wise way to create a state hierarchy. At its coarsest level is a meta-state that subsumes all original graphs $G_i \in DG$.

   For each level of the hierarchy (except for the coarsest one), corresponding meta-transitions are determined. If a meta-state contains a graph $G_i \in DG$ and another state contains the subsequent graph $G_{i+1} \in DG$, then a meta-transition is inserted between the two states.

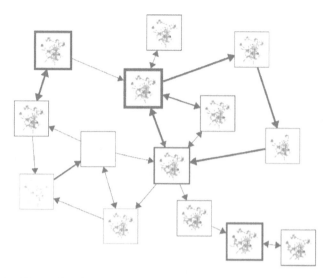

Figure 5.40 Example of a state-transition graph characterizing an underlying dynamic graph. Courtesy of Steffen Hadlak.

Overall, the outlined procedure generates a hierarchical clustering of the original dynamic graph. Each level of the clustering hierarchy corresponds to a state-transition graph that represents the dynamic graph at a different level of abstraction.

To support the visual analysis of large dynamic graphs, a state-transition graph can be visualized as in Figure 5.40. The states are shown as boxes with an embedded miniature node-link diagram that communicates the representative structure of the subsumed original graphs. The border width of the boxes reflects the number of subsumed graphs. Transitions are visualized as directed links between states. Link width encodes the frequency of state changes subsumed by a transition. This visual design allows us to detect interesting temporal patterns in a dynamic graph:

- *Dominant graph structures* are visible as states whose boxes have wider borders.

- *Rare graph structures* manifest in states with thin borders and only few thin incident transitions.

- *Typical structural changes* are expressed by paths along transitions that have wider links.

- *Recurring structural changes* correspond to cycles in the state-transition graph.

- *Branching behavior* is indicated by states having more than one outgoing transition.

Demonstrating Example

Let us next demonstrate the benefit of the introduced clustering methods for the visual analysis of multivariate dynamic graphs. Our example concerns the analysis of link quality in a wireless network operated by the OpenNet community. The network consists of 297 WiFi devices and 2,008 links between them monitored at 217,253 time points (five months at minute resolution). Each link has a quality attribute that describes the probability of a successful packet transmission along the link. The goal of OpenNet is to provide an overall satisfying network quality. Visually analyzing the dynamically changing network allows them to identify, locate, and quickly resolve quality issues.

To support the analysis, both attribute-based and structure-based clustering can be utilized. Attribute-based clustering groups the devices for which the average link quality of their incident links evolves similarly. The clustered supergraph enables OpenNet to distinguish clusters in the network with varying, consistently high, or consistently low link quality. Analyzing the derived time series at multiple temporal scales as in Section 5.3.2 can provide more insight into the evolution of the link quality. Coarser temporal scales facilitate the exploration of global trends, whereas finer scales support the inspection of more fine-grained events such as dropouts.

Dropouts describe incidents that may affect larger parts of the network. They are critical if they occur frequently. To investigate this aspect, structure-based clustering is applied and a generated state-transition graph is visualized in Figure 5.41a. Wide borders of the states emphasize recurring structures, which have a high impact on the stability of the wireless connections. As we can see, S_1, S_2, and S_3 represent such frequent states.

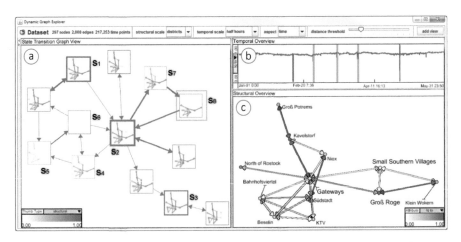

Figure 5.41 Analyzing a wireless network supported by structure-based clustering. (a) State-transition graph; (b) Average link quality of selected state; (c) Representative graph structure of selected state. Courtesy of Steffen Hadlak.

For a more in-depth analysis of states, two additional views are available, a temporal view and a structural overview. They depict further information for a selected state. The temporal view in Figure 5.41b visualizes the average quality of all links contained in the selected state. The structural overview in Figure 5.41c shows the state's representative graph structure. When the structural overview is well connected, and the temporal overview shows a sufficiently high link quality without dropouts, then we can conclude that the state is sufficiently functional. By analyzing all states in this manner, it is possible to detect the states with dropouts. Whether these states are singular events or occur more frequently can be seen from the border width of the state's box in the view of the state-transition graph.

This small example demonstrates how the theoretical concepts for clustering multivariate dynamic graphs can be employed in practice to address a real-world data analysis problem. The clustering not only helps us to reduce the amount of data to be visualized, but also to detect characteristic structures and interesting temporal patterns.

At this point, we conclude the section on grouping data elements with the help of classification and clustering. We have learned that classification subdivides the data space in order to define meaningful subspaces, the classes. Each class describes data with particular properties. How to define the classes depends on the application domain and the tasks to be performed.

Clustering, on the other hand, creates groups based on the similarity of data elements. The clustering outcome depends not so much on domain-specific or task-specific constraints, but primarily on the data and on the configuration of the clustering. The latter includes the selection of appropriate similarity measures and suitable clustering strategies, the parameterization of the selected methods, the weighting of data attributes, and the treatment of outliers. All these decisions lead to different clustering results.

Overall, configuring classification and clustering approaches is a non-trivial task. Interactive visual representations can help us obtain suitable setups that reliably extract and convey the key characteristics of the analyzed data. Only then can classification and clustering support the generation of meaningful overviews that *show the important* as suggested by Keim's visual analytics mantra [Kei+06].

5.5   REDUCING DIMENSIONALITY

Clustering and classification as discussed before are concerned with *data elements*. In this section, we will reduce the number of *data variables*. Data with very many variables are difficult to analyze for the following reasons:

- Data issues: With many variables, the data space gets so huge that its data elements become sparse. As a direct consequence, the number of data elements needed to estimate a function describing a data property grows exponentially. With the sparsity of the data space, the data get more uniform and global structures and features disappear. This phenomenon has been coined as the *curse of dimensionality* [Bel61].

- Visualization issues: With an increasing number of variables, the visual representations become more and more complex. Over-plotting can become a serious issue. It is more difficult to perceive relevant structures, outliers, or trends.

- Performance issues: Visualizing many variables also places higher demands on memory and runtime. It becomes more and more difficult to ensure interactive frame rates.

In order to address these issues, dimensionality reduction aims to reduce the number of data variables. The goal is to figure out which data variables carry relevant information and to focus the data analysis on them, neglecting the unimportant data variables. This can substantially simplify the analysis.

The idea of dimensionality reduction is to project data elements from the high-dimensional data space onto a lower-dimensional projection space in such a way that the original information is preserved as much as possible. There are different approaches to perform the projection. Their difference lies in how they measure which parts and characteristics of the data are considered relevant and hence should be preserved in the projection space.

Two different approaches can be differentiated: linear methods and non-linear methods. *Linear methods* define the dimensions of the projection space as linear combination of data variables. A prominent linear method is the *principal component analysis* (PCA). It is based on a rotation of the data space so that the dimensions of the projection space, the *principal components*, describe the directions of the largest data variations. In contrast, *non-linear methods* define the dimensions of the projection space with regard to certain proximity constraints. For example, with *multi-dimensional scaling* (MDS) pairwise data distances are preserved.

It can be quite difficult to figure out which approach to apply and how to configure it. As before, we can employ interactive visual methods to help users understand and steer the automatic computations toward meaningful projections. Next, we discuss this in more detail for the example of using PCA to determine relevant data variables.

5.5.1 Principal Component Analysis

In order to get a better understanding of PCA, let us briefly explain the general idea by the simple example in Figure 5.42. Figure 5.42a depicts a number of data elements in a two-dimensional data space spanned by the variables V_1 and V_2. The figure also shows the two principal components PC_1 and PC_2. We can easily see that PC_1 is aligned with the axis of the greatest variance in the original data. PC_2 is orthogonal to PC_1.

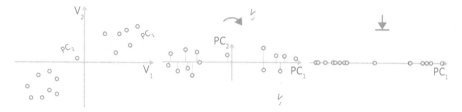

(a) Original data space.   (b) Principal component space.   (c) Reduced space.

Figure 5.42   Reducing dimensionality with principal component analysis.

In Figure 5.42b, the data space has been rotated according to PC_1 and PC_2 to define the projection space, more specifically, the principal component space, or short PC space. In PC space, most of the information inherent in the data elements is distributed along the horizontal axis, whereas the vertical axis carries only little information. This allows us to apply dimensionality reduction and to represent the data just by PC_1.

Figure 5.42c illustrates the result of the dimensionality reduction. The data elements are now only given with respect to PC_1. Although we lose the information with respect to PC_2, the one-dimensional representation still clearly shows two data groups at the beginning and the end of PC_1 and one outlier in the center. This simple example shows us how PCA can be utilized to decrease the data complexity, while still maintaining the distribution of data elements along the major data trend described by PC_1.

For actually calculating the principal components, we need tabular data modeled as an $n \times m$ matrix X, where the columns represent m data variables and the rows represent n data elements. By applying a *singular value decomposition*, the matrix X is decomposed as follows:

$$X = W \cdot \Sigma \cdot C^T$$

The rows of the matrix C^T are the transposed eigenvectors of the data's covariance matrix $X^T X$. They define the principal components as linear combinations of the original data variables. The factors involved in the linear combinations are denoted as *loadings*. The loadings capture how much an original data variable contributes to a principal component.

The matrix Σ is a diagonal matrix. The diagonal entries contain *significance* values resulting from the ranked square roots of the eigenvalues of the data's covariance matrix $X^T X$. A principal component's significance tells us how much original data variance it covers.

The principal components in C^T are ordered with regard to their significance. The first principal component captures most of the data's variance and thus represents the major data trend. The second principal component captures most of the remaining variance and so on. The least significant principal components, that is, the last rows of C^T represent only very small variances. These components can be neglected in order to reduce dimensionality.

The rows of the matrix W contain the transformed coordinates of the data elements in the principal component space. These coordinates are denoted as *scores* to make them distinguishable from the original positions of the data elements.

Next, we will see how principal components, loadings, significance, and scores can be utilized for data analysis purposes.

5.5.2 Visual Data Analysis with Principal Components

Visual data analysis with principal components typically requires two mutually dependent analysis perspectives. First, we investigate how visual methods can help us understand and steer the dimensionality reduction with PCA. Second, we describe how the actual data analysis can be carried out based on the principal components.

Visual Support for Principal Component Analysis

Before the PCA can be calculated, it needs to be decided what should be included in the data matrix X. Theoretically, X can cover the whole dataset. However, it can make sense to compute principal components only for a data subset depending on the task at hand. For example, variables describing time and space are often excluded to leave them as a frame of reference. Moreover, highly correlated variables might overemphasize certain data trends, and outliers could distort the analysis results. It is up to the user to exclude these variables and data elements from the PCA.

The decision regarding which parts of the data should be excluded can be supported by visual methods. Table-based visualizations as introduced in Section 3.2.1 of Chapter 3 can be used for this purpose. When a table-based visualization is sorted properly, highly correlated variables and outliers to be ignored can be recognized easily.

Once the PCA has been carried out on the defined data subset, we obtain the principal components. For each principal component, we know its significance and how each variable contributes to it, that is, the loadings. Based on this information, we can reduce the PC space. Typically, only the two or three most significant principal components are retained, because they capture the major data trends and can easily be plotted in 2D or 3D visual representation. However, less significant principal components might also bear interesting and unexpected information. This raises the question of how to choose principal components in an adequate way.

To answer this question, we need to take not only the significance but also the loadings into account. In fact a subset of principal components sufficiently characterizes the original data, only if:

1. all loadings of these principal components are sufficiently high, that is, all variables are represented well, and

2. the loadings of the remaining principal components weighted by their significance are sufficiently small.

Again, a table-based visualization can help us check whether these two conditions are satisfied [MNS06]. In Figure 5.43, we illustrate this for a demographic dataset. The rows of the table show the variables of the original data, including population, literacy, or life expectancy. The columns contain the principal components ordered from left to right according to their significance.

The table cells visualize loadings weighted by significance. Each table cell corresponds to a pair of a data variable and a principal component. Per cell, a bar encodes information as follows. The bar extends from the cell center to the right and is colored blue when the loading has a positive sign. The bar extends to the left and is yellow when the sign is negative. The length of the bar represents the value of the variable's loading weighted by the principal component's significance. As such, the bars in a column act as an indicator for the relevance of the corresponding principal component. The bars in a row show the influence of the data variables on the different principal components.

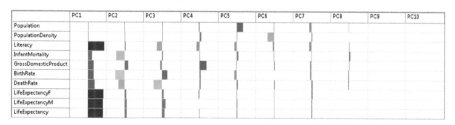

Figure 5.43 Table-based visualization of loadings weighted by significance. Reprinted from [Aig+11].

In Figure 5.43, we can see many longer bars in the first three columns. This means that the first three principal components actually capture a lot of information, suggesting that all other components could be candidates for reduction. However, the first two variables, population and population density, hardly contribute to the first three principal components. Hence, condition one as stated above is violated. Moreover, there are also a couple of longer bars in columns three to six, which means that also the second condition is not met. If we retain only the first three principal components, we will likely lose information. Instead, we should consider at least four, better five or six principal components. Only then are all variables faithfully represented by the reduced PC space.

However, still interesting information might be hidden in the data. In order to reveal them, it may help to look at the unweighted loadings. In Figure 5.44, the plain loadings are visualized in the same way as before. Now, we can

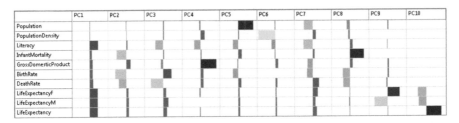

Figure 5.44 Visualizing unweighted loadings empasizes the contribution of individual variables to each principal component. Reprinted from [Aig+11].

see the contribution of the individual variables to each principal component. This facilitates the detection of deviating data behavior. For example, in the ninth principal component, we can identify a difference between the loadings of the two variables for life expectancy of females and males. This difference might suggest including the principal component in the analysis as well or investigating the two variables separately in more detail.

In summary, the above examples demonstrated how the visualization of loadings and significance can help us make an informed decision when it comes to reducing dimensionality of data. Next, we briefly describe how the actually reduced principal component space can be analyzed.

Visual Data Analysis Based on Principal Components

Visual data analysis based on principal components means showing the data elements in the PC space, more precisely, in the reduced PC space. Generally, the visualization is straightforward. The data elements can simply be visualized according to the *scores* calculated during the PCA. In general, the visualization techniques for multivariate data introduced in Section 3.2 in Chapter 3 can be utilized for this purpose. For example, we could use parallel coordinates with as many axes as we have principal components and plot each data element based on its score values. Alternatively, we could adapt the idea of the tabular visualization used before. We maintain the principal components in the columns, but show data elements as rows. Each table cell would then represent a score value. While these are all technically feasible visual representations, a key issue needs to be addressed.

Because the principal components spanning the PC space correspond to a combination of several original data variables, interpreting the scores and relating them back to the original data space is difficult for a human observer. Therefore, we need visual support helping users gain an understanding of the rather abstract principal components. Two visual approaches can be useful in this regard:

- *Annotating principal components with data variables*: The principal components are labeled with the names of those variables that mainly contribute to them. To avoid clutter, labels can be shown on demand when the user hovers over a principal component with the cursor.

- *Associating scores with data values*: A subset of scores in PC space can be associated with their corresponding values in data space. This can be done either by a side-by-side visualization of scores and data values or by a combined visualization. In the side-by-side case, brushing & linking allows users to connect the scores from one view with the data values of the other view. In the case of a combined visualization, a limited number of principal components and data variables are presented simultaneously. An example could be our table-based visualization. As before, the table columns represent the principal components, but the rows now show the original data variables, and the bars in the cells represent scores. In such a visual representation, small bars symbolize data values that follow the main trend of a principal component, whereas larger bars correspond to substantial deviations from the trend.

To conclude, again we have seen that automatic computations, PCA in this case, can support the interactive visual data analysis. PCA is a helpful approach to determine which data variables play an important role and which can potentially be neglected. Knowing this, we can simplify the visualization and concentrate the analysis on those parts of the data where insight are likely to be made.

Yet, we have also seen again that employing automatic computations requires a careful configuration and a good understanding of the consequences. Visual methods can help users in this regard. In fact, we need a tight interplay of automatic calculation and human sense-making based on visual representations.

5.6 SUMMARY

In this chapter, we discussed automatic computational methods to support interactive visual data analysis. The key idea was to reduce the complexity of the data and their visual representations. A figurative overview of the covered methods is provided in Figure 5.45.

First, we briefly described how the complexity of visual representations can be addressed with density-based visualization and bundling. The idea of density-based visualization is to compute and visualize the frequency of data elements. Bundling improves the visual structure in visualization images by summarizing graphical elements.

For the most part of this chapter, we studied ways to reduce the size and complexity of the data. A first strategy was to focus the analysis on relevant parts of the data. The degree-of-interest (DoI) approach was introduced as a means to determine the relevancy of data elements based on well-defined interest functions. A related approach is feature-based visual analysis, where characteristic data features are automatically extracted and visualized.

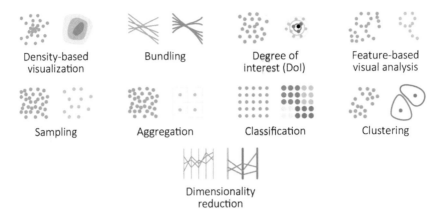

Figure 5.45   Overview of automatic computational methods to support interactive visual data analysis by reducing the complexity of the data and their visual representations.

We further discussed how abstracting and grouping data elements can facilitate the data analysis. Sampling and aggregation are two well-established data abstraction methods. Sampling methods aim to generate a reduced, but still representative subset of the original data. Aggregation methods replace the data values within an interval by statistical descriptions such as the average or the mode. Classification and clustering are typical strategies for grouping data elements. With classification, the data space is partitioned into classes with specific properties, whereas with clustering, the data are organized in clusters of similar data elements. Last but not least, we described how dimensionality reduction can help us reduce the complexity of data with many variables.

All of the above approaches address the "Analyse First"-step of the visual analytics mantra as stated in the beginning of this chapter [Kei+06]. Although each approach pursues a different strategy, they all have in common that they transform unstructured raw data into a form that emphasizes the data's basic properties and key characteristics. This way, the data is prepared to be more meaningful for human exploration and insight discovery [Sac+14].

However, applying automatic computations typically requires a careful configuration. What computational methods should be applied, maybe even in combination, and how should their parameters be set? Finding suitable answers to these questions is important, because different methods and different parameter settings lead to different results, and not all of them might be appropriate for the task at hand.

In this chapter, we have seen several examples where visual methods were employed not only to show the data, but also to help users understand how automatic computations work and how their parameters should be set. In fact, we have seen a mutual relationship between automatic computation and visual methods. On the one hand, automatic computations support the visual analysis by extracting important characteristics of the data. On the other hand, visual methods help us to configure the automatic computations appropriately. Both sides of the relationship are tied together by the human user who has interactive control over all involved methods [End+17]. This control paired with expressive visual representations enhances trust in and interpretability of the generated analysis results.

Overall, we have now dealt with the fundamental ingredients of interactive visual data analysis. In the next chapter, we will move on to advanced concepts for visualization, interaction, and automatic computation.

FURTHER READING

General Literature: [Kei+06] ● [Kei+10] ● [End+17]

Decluttering Visual Representations: [NH06] ● [BW08a] ● [LHT17]

Focusing on Relevant Data: [RPS01] ● [DGH03] ● [Abe+14]

Abstracting Data: [ED06] ● [EF10] ● [Lub+12]

Grouping Data: [HKP11] ● [XW05] ● [Emm+16]

Reducing Dimensionality: [Jol02] ● [LV07] ● [Sac+17]

Advanced Concepts

CONTENTS

I N THE PREVIOUS Chapters 3 to 5, we described the three fundamental parts of interactive visual data analysis: the visualization, the interaction, and the automatic analysis support. We learned about basic visualization strategies, interaction techniques, and computational methods, which usually work in concert to facilitate data analysis activities. The knowledge from the previous chapters enables the reader to apply existing approaches to analytic problems at hand or to design new ones if necessary.

The goal of this chapter is to broaden our view on interactive visual data analysis to advanced topics that are not yet mainstream, but have the potential to become so in the future. Following the structure of the previous three chapters, we will discuss three selected topics:

- advanced visualization in multi-display environments,

- advanced interaction through user guidance, and

- advanced automatic computation via progressive procedures.

Section 6.1 will discuss the visualization of data in multi-display environments. We will see that such environments provide us with interesting opportunities to enhance the visual data analysis. They allow us to show more visual content and also to show it to more people. This enables collaborative data analysis and in-depth discussions involving several experts.

Section 6.2 deals with facilitating the human-computer partnership by providing guidance that helps users carry out expedient analysis actions. We will discuss a conceptual framework that characterizes guidance in the context of interactive visual data analysis, and two selected examples will illustrate how guidance can actually be implemented.

Finally, in Section 6.3, we examine the challenge of time-consuming automatic computations, which may impede the data analysis. One answer to this problem is to progressively compute intermediate results and present them to the user timely, allowing for early evaluation and, if necessary, intervention.

With these three topics, we are moving into a terrain of active data analysis research. That said, the content to come is naturally subject to further academic discussions and revisions. Still the following sections provide an outlook on exiting new perspectives on interactive visual data analysis. Let us start with visualization in multi-display environments.

6.1 VISUALIZATION IN MULTI-DISPLAY ENVIRONMENTS

Most of the visual analysis solutions described in this book are designed for traditional environments where a single user is working with one or two displays. However, such environments are limited in two regards. First, only a single individual is involved in the data analysis. Critical reflections of results or creative discussions of alternative analysis strategies are hardly possible. Second, the available display space is limited. This can make the analysis of larger volumes of data difficult.

A natural step to address these limitations is to bring the visual data analysis to advanced *multi-display environments* (MDEs). MDEs can facilitate visual data analysis by enabling more users to observe more data. The increased overall display space not only makes it possible to visualize more data elements, but also to show more aspects of the data simultaneously. The increased physical size of MDEs allows multiple users to study the visualized data, which promotes collaborative analytic work.

Collaborative analysis in MDEs is particularly useful in scenarios where experts from different application domains have to discuss heterogeneous data to arrive at a common understanding of complex phenomena. One such scenario is the study of the impact of climate change. Climate change affects a multitude of sectors, such as agriculture, forestry, ecosystems, and economy. Experts from many fields need to work together and share their domain-specific observations and data analyses to better understand the crucial questions related to climate change.

Collaborative visual data analysis in MDEs can be a paradigm shift away from the otherwise single-user setting at regular desktop computers. Yet, this shift also comes with new challenges. For example, distributing and

Figure 6.1 Smart meeting room at the University of Rostock. Reprinted from [ENS15].

arranging visual representations manually on multiple displays can be a time-consuming task that should not be burdened on the users. Keeping track of analysis steps and intermediate findings is also more difficult in a collaborative setting. Therefore, the visual data analysis in MDEs must be facilitated by dedicated methods that relieve users of laborious manual work and allow them to concentrate on their analytic objectives. This section outlines how advanced visual data analysis can be implemented in MDEs.

6.1.1 Environment and Requirements

To start with, we will briefly introduce a particular MDE and the requirements that need to be considered to support collaborative visual data analyses in it.

Environment Our environment is a smart meeting room as illustrated in Figure 6.1. A smart meeting room is an instance of *smart environments*, which Cook and Das define as *"a small world where all kinds of smart devices are continuously working to make inhabitants' lives more comfortable"* [CD04]. The smart meeting room combines several heterogeneous displays to form a coherent display space in which content can be distributed even across device boundaries. Moreover, the smart meeting room can dynamically integrate various input, computing, and output devices into the device ensemble, which allows users to bring and use their own personal devices. These characteristics make a smart meeting room a distinguished MDE, namely a *smart* MDE. While the MDE provides a technical basis, it is the MDE's smartness that enables the actual visual data analysis.

Requirements for Collaborative Visual Analysis In order to support the collaborative visual data analysis in smart MDEs, we must understand how users work. In the first place, the users compile a corpus of information to be analyzed and discussed. Any user may contribute contents, such as visualization images, slides, or documents. For simplicity, let us call these different contents just *views*. Then, the users define which views should be shown together and in what sequence. This way, a kind of presentation is created that meaningfully aligns the views with the analysis goals. While setting up the presentation is the task of the users, distributing and arranging the views on the multiple displays should be the task of the machine. This brings us to the first two requirements:

- The system should afford easy contribution of views and creation of the presentation. This ensures that the users can control the content and the flow toward the analysis goal.

- The system should distribute and lay out the views in the environment automatically according to the state of the presentation. This relieves the user from cumbersome manual layout tasks.

While discussing and analyzing the data contained in the views, it is often necessary to go back to previously shown information, to add, adjust, or replace certain visual representations, or to compare performed analysis steps. Furthermore, newly gained insights may raise new questions, which in turn might require inspecting additional data. In this case, the users must be able to generate the needed visualizations without interrupting the ongoing presentation. From this, two further requirements can be derived:

- The system should support the seamless switching between the presentation of views and the generation and adjustment of views. This narrows the gap between phases of presentation and phases of more personal exploratory analytical work.

- The system should keep track of the analysis history, including changes to the presentation and the findings made during the analysis session. This helps users to return to previously discussed aspects and to recap the analysis and its results.

Next, we will illustrate by concrete methods how these requirements can be addressed to support the collaborative visual data analysis in smart MDEs.

6.1.2 Supporting Collaborative Visual Data Analysis

The solutions to be presented next focus on supporting four key tasks: the creation of the presentation, the layout of views on multiple displays, the adjustment of presentation and views during the analysis, and the recording of the visual analysis process.

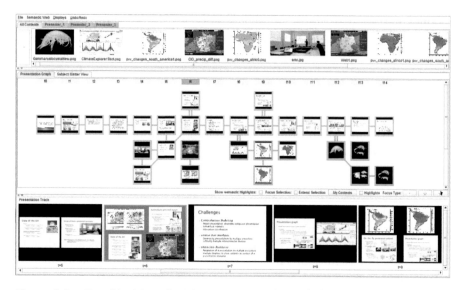

Figure 6.2　Graphical interface for creating and controlling multi-display visual analysis presentations, including content pool (top), logical presentation structure (middle), and preview (bottom). Courtesy of Christian Eichner.

Creating a Multi-display Presentation

The creation of presentations for collaborative visual data analysis can be supported by a graphical user interface as shown in Figure 6.2. The interface is available on all devices being connected to the smart MDE, so that everyone can participate in preparing the presentation, even from their personal devices.

The top panel of Figure 6.2 represents the content pool and affords the collection of views. Visualization images, slides, or documents can be added to the content pool by simple drag and drop. A special thing about the collected views is that they can be linked to the software that was used to generate them. The linked software can be started at any time by clicking the corresponding view. This allows users to alter existing views or to generate new ones on the fly during the presentation.

The middle panel in Figure 6.2 serves to set up the presentation based on the views in the content pool. For this purpose, views from the top panel are dragged to the vertical layers in the middle panel. From left to right, the layers represent the sequence of views to be shown in the course of the presentation. Views within the same layer are visible at the same time. Links can be added between views to define spatial and temporal constraints for the later automatic layout of the views on multiple displays. Spatial constraints (within layers) link views that are to be shown on the same display, whereas temporal constraints (between subsequent layers) link views to be replaced when advancing the presentation from one step to the next.

The bottom panel in Figure 6.2 provides a preview of the presentation. It tells the users which views are currently being shown in the smart MDE, which views were displayed before, and which views are still to come. This panel is also used to advance the presentation to the next layer, to return to previously shown layers, or to directly go to a certain layer.

In summary, the described graphical interface enables users to set up a multi-display visual analysis presentation. To avoid conflicts during the collaborative work and to ensure a consistent presentation, a moderator should chair the process.

Layout of Views on Multiple Displays

Smart MDEs provide the technological basis for distributing the graphical contents on multiple displays. What is needed to support collaborative visual analysis is a mechanism that distributes and arranges the views according to the prepared presentation. This can be done with a dedicated multi-display layout algorithm.

The automatic view layout has to be carried out in two steps as illustrated in Figure 6.3. In the first step, views are allocated to the available displays of the environment. The second step computes the arrangement of views per display. Because both steps must consider various influencing factors, including the number and the properties of the available displays and the user-defined temporal and spatial constraints, it makes sense to define the automatic layout as an optimization problem [Eic+15]. The goal of the optimization is to find view positions such that the overall layout quality is maximal.

Three terms contribute to the overall layout quality: the spatial quality, the temporal quality, and the visibility quality. The spatial quality is high if the views being linked via spatial constraints are near to each other, ideally on the same display. The temporal quality is high if temporally stable layouts are produced. That is, views being linked via temporal constraints are ideally presented at the same position when advancing the presentation from one step to the next. The visibility quality is a bit more complex to define. It rates how well users can see the different views in the display environment, which can

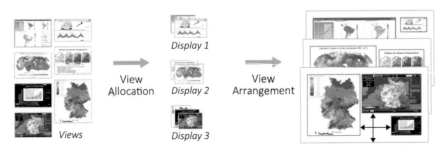

Figure 6.3 Basic two-step procedure of the automatic view layout.

be approximated by considering the directional visibility of views. A smart MDE can estimate this information based on the display configuration (size, position, and orientation of displays) and by tracking the participating users (position and viewing direction of users) [RLS11].

The outlined optimization problem has to be solved whenever a change happens either to the presentation or to the environment. In order to manage to compute the layout for a dozen of views in less than a second, a heuristic optimization approach should be applied. For example, by employing the *branch-and-cut* algorithm, overestimating the achievable qualities of possible layouts, and reusing previous calculations, it is possible to obtain expedient results in a short time.

Interactive Adjustments in Multi-display Environments

Once the layout mechanism has delivered its initial results, the views are shown in the smart MDE and the data analysis can start. The regular process will be to discuss the depicted information, agree on intermediate findings and results, and advance the presentation further. Yet, with changing topics of interest, the actual course of the analysis might divert from the originally planned one. In this case, the users must be able to adjust the presentation on the fly. To this end, two types of interactive adjustments should be supported: adjustment of the layout of views and adjustment of the content of views.

Adjusting the Layout of Views   Adjusting the view layout involves moving and resizing views. For example, views might need to be moved from one display to another for side-by-side comparison. Fine details spotted during the comparative analysis could make it necessary to enlarge a view. To this end, the moderator (or another authorized user) can select and adjust a certain view to improve the visibility of details.

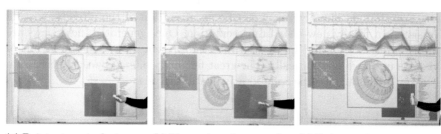

(a) Point at central view.   (b) Move view downward.   (c) Enlarge view.

Figure 6.4   Adjusting position and size of a view using a Wii Remote controller. Reprinted from [Rad+12].

An example of such a layout adjustment is illustrated in Figure 6.4. We already saw these visual representations of graphs in Section 3.5.3 of Chapter 3, but now they are projected onto a canvas of the smart MDE. A user points at

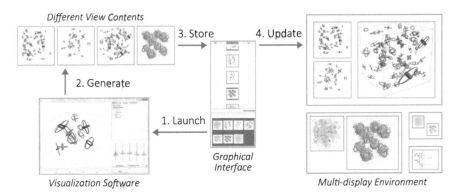

Figure 6.5 Changing the content of views by launching visualization software.

the central Magic Eye View using a Wii Remote controller, moves the view a bit downward, and then enlarges the view to make it stand out. On the user's side, these interactions are easy to perform. On the system's side, dedicated mechanisms such as device integration and interaction mapping are employed to facilitate the interaction [Rad+15].

Adjusting the Content of Views We already mentioned that views can be linked with a compatible software. This makes it possible to re-generate the content of views to better align them with varying analysis requirements. Figure 6.5 illustrates an example with the feature-based visualization already described in Section 5.2.2.

First, the linked visualization software is launched from the graphical interface with a click on the view's thumbnail. This can be done by users who have the visualization software installed on their personal device. On the personal device, the data can then be explored until a suitable new view has been found. The new view can be an alteration of the already existing version or a totally new visual representation. Once generated, new views are immediately stored and integrated into the presentation, and the layout of views is updated automatically. The great flexibility offered by this mechanism is a key advantage for the collaborative data analysis.

Keeping Track of Collaborative Data Analyses

It is generally accepted that keeping an analysis history is beneficial for coordinating the insight-generation process and reflecting about it [KNS04]. Commonly, analysis histories store information about the interactive adjustments performed by users and the findings that have been derived. In a collaborative multi-display setting, it makes sense to log additional information about who contributed and changed views and when and where views have been displayed. Some of this information can be determined automatically. For example, the

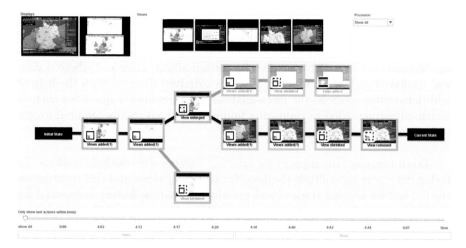

Figure 6.6   Graphical interface for analysis coordination and meta-analysis, including filtering support (top), analysis history graph (middle), and timeline with undo and redo buttons (bottom). Courtesy of Christian Eichner.

view layout mechanism can keep track of when and where views were shown. Interactive adjustments can also be recorded automatically, including the actions that were taken, the users who carried them out, and the resulting visual representations. Yet, some information cannot be derived automatically. For example, findings derived from a view need to be annotated manually.

The recorded and annotated information is stored in the form of a graph. The graph's nodes represent views, more precisely, the state changes logged per view. As such, a node captures a piece of analytical progress made during the data analysis. Links between nodes form paths of analytical progress as defined by the sequence of actions taken.

To actually gain from the analysis history, it can be displayed in a graphical interface as illustrated with a small example in Figure 6.6. During an ongoing analysis session, the interface can be employed to reset the analysis to a previous state via selective undo and redo. "Selective" means it only affects operations that were triggered by a certain user, affected a specific view, or concerned a particular display. This is helpful when the data analysis stalls in a dead end or if the participating users cannot come to an agreement about findings and intermediate analysis results. Undo and redo allows the moderator to keep the analysis going, for example, by collecting further evidence for or against a hypothesis from previous views. If, after returning to a previous state, an alternative course of actions is pursued, a new analysis branch is created, which is also visible in Figure 6.6.

A graphical depiction of the analysis history can also support a post-hoc meta-analysis to understand how individual analysis steps contributed to the generation of new insights. For example, in Figure 6.6, we can see that three alternative analysis routes were tried out. Small icons overlaid on the thumbnails indicate which interactions were performed, and the thumbnails' colored borders tell us who performed them. When a thumbnail is clicked, an on-demand text box will provide information about when and where a view was displayed and which findings have been derived from it. With the help of additional filter controls, the meta-analysis can even answer questions such as which adjustments led to promising findings or which results required longer discussions.

Taken together, the support for creating collaborative analysis sessions, for laying out views on multiple displays, for adjusting views and their content on the fly, and for keeping track of and utilizing an analysis history is essential for facilitating visual data analysis in smart MDEs. Next, we will illustrate how a smart MDE can be put to use for analyzing the impact of climate change.

6.1.3 Multi-display Analysis of Climate Change Impact

As already indicated, analyzing the impact of climate change requires the collaboration of multiple experts. In this section, we sketch a scenario with experts from meteorology, forestry, and hydrology. The course of collaborative analysis could be as follows [Eic+15].

First, the meteorologist wants to explain extreme precipitation events based on visual representations that she already has on her laptop computer. She logs into the smart MDE and contributes her contents via the graphical interface introduced earlier. The automatic layout will make sure that all experts can see the visual representations well. Now the joint analysis can start.

The explanations of the meteorologist trigger the hydrologist to join the discussion. To better make his point, the hydrologist uses the graphical interface on his computer to put a visualization about groundwater recharge into the content pool. Then, he defines a spatial constraint to link his view to a precipitation view of the climatologist. In the blink of an eye, the view layout in the smart MDE is automatically updated to show the linked views side-by-side on the same display. This allows a comparison and in-depth discussion of the views.

At some point, the experts realize that they need additional views to make progress. Therefore, the hydrologist connects his view with the climateimpactsonline.com web portal. This allows him to generate the necessary views on the fly. Depending on the discussion of the new views, the experts can decide to either keep them or to perform undo operations to reset the presentation to the original views.

Finally, the forestry expert takes part in the discussion. She explains how climate-related risks in Africa and South America are expected to develop in comparison to Europe. To illustrate this, she puts further views into the content pool and defines spatial and temporal constraints to connect them to the existing views. Again, the presentation is updated automatically.

In the course of the visual analysis, the knowledge and findings of the different experts are discussed and combined to form a broader understanding of the impacts of climate change. After the analysis session, all experts can inspect the analysis history to reflect about the discussion or reproduce the analysis results.

In summary, we see that bringing interactive visual data analysis to smart multi-displays environments is an exciting opportunity for collaborative sense-making. Yet, in order to fully exploit this opportunity, it is necessary to design dedicated support to facilitate the analysis. Here, we illustrated selected solutions based on a mix of automatic methods and interactive graphical interfaces implemented in a smart multi-display environment, which together form an advanced visualization environment.

Next up in this chapter on advanced concepts is advanced interaction through user guidance.

6.2   GUIDING THE USER

The meta-analysis in the previous section confirmed what we already stated several times: Visual data analysis is not a one-way road but a dynamic process during which several what-if scenarios are tried out.

How should the data be processed with analytical calculations? Should clusters be computed, if yes, how many clusters are expected? Are certain data variables to be excluded from the processing? Which techniques should be employed to visualize the data? Should different data facets be represented in an integrated fashion or in separate dedicated views? Which parts of the data should be explored in what sequence? Only if appropriate answers to these questions can be found is it possible to make progress toward the desired analytic results.

But who has to answer the questions, the users? Or could the machine step in and help us out in certain situations? This is what *guidance* is about [Sch+13b; Cen+17; Col+18]. As a response to the challenge of ensuring analytic progress, it is the goal to guide users towards choices that present the most interesting aspects of the data with the most suitable combination of visual, interactive, and analysis methods.

In Section 5.3.2 of Chapter 5, we already saw how guidance can assist users when exploring multi-scale time series. We will next discuss the relatively new topic of guidance in more detail. The first part of this section will be of conceptual nature offering an in-depth characterization of guidance. The second and the third parts will be more practical. Two examples will illustrate how guidance can help users in analyzing complex data other than time series.

6.2.1   Characterization of Guidance

So, what is guidance? In order to make more clear what we mean by guidance, we will start with a definition of guidance in the context of interactive visual data analysis, discuss its key aspects, and then introduce a conceptual model.

Definition of Guidance

Guidance is a broad term with much room for interpretation. The Oxford Dictionary and the Merriam-Webster Dictionary define guidance as "advice or information aimed at resolving a problem or difficulty" and "the act or process of guiding someone or something". These definitions are interesting because they highlight guidance as a *process* aiming at solving a *problem*. This is also reflected in the following definition of guidance in the context of interactive visual data analysis [Cen+17]:

"Guidance is a computer-assisted *process* that aims to actively resolve a *knowledge gap* encountered by users during an *interactive* visual analytics session."

Ceneda et al., 2017

The three important aspects of this definition are emphasized in italics. First, guidance is a dynamic *process* that runs alongside the regular data analysis activities of the user. Second, there is a *knowledge gap* that causes the data analysis to stall. The user does not know how to proceed. The goal of guidance is to narrow the knowledge gap. Finally, the definition of guidance describes an *interactive* scenario where human and machine cooperate.

The above definition also suggests what guidance is not. Guidance is not just an additional algorithm that computes a unique answer to the knowledge gap. Typically, this is not even possible due to ill-defined or too complex analytic problems. If guidance were able to compute a precise answer, we could neglect the interactive visual approach to data analysis at all, compute the answer, and provide it to the user right away. But this would contradict with the idea of having the human in the loop.

That said, guidance does not take over the reasoning part. Instead, guidance facilitates the data analysis to help users in forming decisions. Making the decisions remains the responsibility of the user. In this sense, guidance is comparable to mentors helping students. While mentors do not necessarily know the solution of the students' problems, they can provide hints as how to approach the problems, guiding the students towards finding solutions on their own.

Figure 6.7   A knowledge gap exists when target or path is unknown.

Aspects of Guidance

All in all, guidance is in fact a catalyst for human-computer cooperation. There are four main aspects that are worth further detailed consideration [Cen+17]:

- **Knowledge Gap:** Why is guidance needed?

- **Input:** What information can be utilized for providing guidance?

- **Output:** How is guidance conveyed and how does it look like?

- **Degree:** How much help does guidance provide?

Knowledge Gap   The knowledge gap captures what a user needs to know to make progress. While one can easily imagine many different knowledge gaps, from a conceptual point of view, there are only two distinct *types* of knowledge gaps:

- *Target unknown.* The user does not know the desired result. For example, the user does not know which data features to look at to falsify a hypothesis.

- *Path unknown.* The user does not know how to reach the desired result. For example, the user knows the relevant features, but has no clue how to configure the visualization to reveal them.

Figure 6.7 illustrates the two types of knowledge gaps. Certainly, if both target and path are known to the user, no guidance is needed.

In general, capturing the knowledge gap is difficult. Users may or may not be aware of their knowledge gap. If users are aware of it, they can actively make it known to the system. If not, the system has to infer the knowledge gap, for example, by detecting deviations from domain conventions or long dwell times during exploration. Ultimately, guidance should narrow down the knowledge gap in a dynamic process that eventually converges to zero knowledge gap.

Input   The input subsumes the different sources of information based on which guidance can be generated. In the context of interactive visual data analysis systems, the following inputs can be utilized:

Data input relates to information that is readily available or derivable from the original data, including the raw data themselves, statistical properties of the data, extracted topology, or meta-data.

Domain knowledge input refers to information that is commonly agreed upon in the application domain, such as domain models and conventions, established workflows, or expert systems.

Visualization input captures what the user is actually seeing on the display. It includes both the specification of the visualization transformation as well as the actual visualization images.

User knowledge input corresponds to information that users input to the system or that the system can infer from the user. Examples include annotations, preferences, or user interests.

History input is based on keeping track of the course of analysis sessions by logging interaction steps, employed methods and parameters, intermediate views, or visited parts of the data.

Output   Concerning the output of guidance, there are two aspects to be considered. In the first place, guidance must be *generated*, and secondly, it must be *conveyed* to the user.

Generating Guidance Conceptually, generating guidance can be modeled as a function

$$guidance(g, i) \rightarrow o$$

that takes the knowledge gap g and some input i from the available sources and then computes a suitable output o. Suitable means the output contains pieces of information that alleviate the user's problem. Hence, iterating the function several times should narrow the knowledge gap. Each iteration contributes a variable amount of knowledge, depending on the user's expertise and perceptual and cognitive abilities.

Different knowledge gaps can be tackled with different guidance functions addressing diverse aspects of the data analysis. For example, if the user is unsure about what data should be investigated, guidance could identify interesting data subsets and suggest navigational routes toward them. If the user needs help in structuring the analysis into a series of tasks that match the analysis objective, the system could hint at what to do next. A user who is baffled by the variety of interactive, visual, and analytical methods could be supported by suggesting suitable combinations of techniques and corresponding parametrizations.

The generated output can address the knowledge gap directly or indirectly. For example, if a user has difficulties configuring a clustering algorithm, *direct* guidance could suggest promising parameter values. For the same problem, *indirect* guidance could accentuate interesting sub-structures in the data, whose analysis (note the indirection) could help the user better understand the influence of parameter values.

Conveying Guidance The next step is to convey the guidance output in a way that actually facilitates the data analysis, yet without interfering too much with it. This can involve enhancing perception or inducing impulses in the user to trigger exploratory actions.

It seems natural that guidance is primarily conveyed visually, for example, by adjusting the visualization, providing visual enhancements, or including additional graphical interface elements. We will later see two examples, where visually conveyed guidance is central. Yet, depending on the application context, guidance can also be conveyed via non-visual channels, including sounds or tactile feedback.

Guidance Degree   The guidance degree characterizes the extent to which guidance is required and actually provided. The degree is defined on a scale whose two extremes are labeled "no guidance/full freedom" and "fully guided/no freedom". Apparently, the degree of guidance is inversely proportional to the users' freedom. In general, an effective guidance solution restricts freedom as little as possible, but as much as necessary.

Ideally, the guidance degree is not fixed, but rather resonates with the course of the analysis session. In practice, the actually delivered guidance typically follows one of three characteristic scenarios:

Orienting: Merely orienting the user represents a low degree of guidance. The objective of orienting is to support the user in building and maintaining a mental map. Providing visual cues hinting at potential targets and suitable paths is a common strategy for implementing orientation. Visual overview techniques also provide a kind of orientation.

Directing: Directing represents a moderate degree of guidance. In contrast to orienting, directing emphasizes a certain preference for a future course of action. The system suggests to the user a set of options that lead to promising results. The suggestions may differ in terms of quality and costs. Visual preview techniques can help users make informed decisions for one or the other option.

Prescribing: A rather high degree of guidance is reached when the system prescribes certain analytical steps towards a specified goal. This can be compared to a guided presentation that takes the user through the analysis process. To keep the human in the loop, it is important to visually present the intermediate steps taken by the system and to make the decisions that lead from one step to the next understandable. Of course it must be possible for the user to regain control and continue the analysis on another path or to another target.

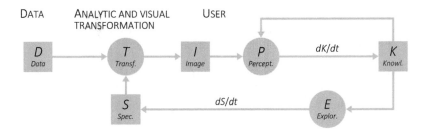

Figure 6.8 Adapted variant of van Wijk's model of visualization. Artifacts as boxes: data [D], specifications [S], visualization images [I], and user knowledge [K]. Functions as circles: analytic and visual transformation (T), perception and cognition (P), and interactive exploration (E). Adapted from [vWij06].

These three scenarios complete our characterization of the different aspects of guidance in the context of interactive visual data analysis. In the following, we will shape the idea of guidance into a conceptual model.

Conceptual Model of Guided Interactive Visual Data Analysis

The conceptual model will help us understand how guidance interoperates with the interactive visual data analysis. To this end, we will attach guidance-related components to an existing model that describes how visually driven data analysis leads to new knowledge. The basic model we will be using is van Wijk's model of visualization [vWij06]. A slightly adapted variant of van Wijk's model is shown in Figure 6.8. It combines the data transformation and knowledge generation models from Sections 2.3.2 and 2.3.3 in Chapter 2 and the human action cycle from Section 4.1.2 in Chapter 4 in a fairly simple and elegant way. Boxes represent artifacts, such as data or images, while circles represent functions that process some input and generate some output.

According to the depicted model, visual data analysis works as follows. Analytic and visual methods transform (T) data [D] into images [I] based on some specifications [S]. Humans perceive (P) the images and cognitively extract the visually encoded information to accumulate more and more knowledge [K]. Based on their accumulated knowledge, users can interactively explore (E) the data by adjusting the specifications. As indicated by dK/dt, knowledge change occurs as a consequence of the interpretation of the visual representations and the exploratory adjustment of the specification dS/dt.

Now let us attach the guidance-related components and see how they can help us to keep the knowledge generation loop going. The central component in the extended model in Figure 6.9 is the guidance generation process (G*). It draws from different sources of input and computes different forms of guidance as output. Before any measures of guidance can be taken, the particular knowledge gap of the user must be known, which is indicated by the link

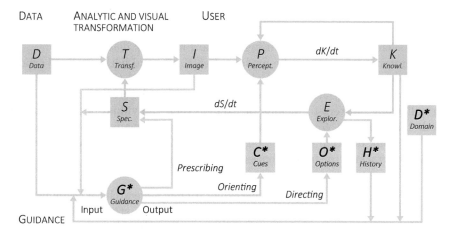

Figure 6.9 Conceptual model of guided interactive visual data analysis. *Added artifacts and functions: domain conventions and models [D*], history and provenance [H*], visual cues [C*], options and alternatives [O*], and guidance generation (G*). Adapted from [Cen+17].

between [K] and (G*). Further sources of input connect to (G*), including the original data [D], the visualization images [I] and the underlying specification [S], the interaction history or provenance [H*], and domain conventions or models [D*].

On the output side, guidance can be delivered in various ways and to different degrees. Basic guidance can show visual cues [C*] alongside the visualization to help users orient themselves. Directing users to promising analysis paths works by determining and offering options [O*] that, if chosen during the interactive exploration, lead to improved visualizations. Finally, the guidance mechanism can take over control and circumvent progress-hindering obstacles automatically by prescribing certain specifications [S] directly.

The sketched conceptual model provides us with a blueprint of how guidance methods are coupled with interactive visual data analysis. Next, we will introduce two examples that illustrate how guidance can be implemented based on the introduced conceptual model.

6.2.2 Guiding the Navigation in Hierarchical Graphs

The first example is about navigational guidance, which we already briefly covered in the introduction in Section 1.2.3. Now, we focus on an approach for guiding the navigation in hierarchical graphs [GST13].

Data, Analytic and Visual Transformation, Exploration A hierarchical graph is a regular graph on top of which a hierarchy defines a nested structure of clusters. Cuts through the hierarchy define views of the underlying graph at different levels of abstraction. The views correspond to regular plain graphs and as such can be visualized using standard node-link diagrams.

Hierarchical graphs can be navigated vertically and horizontally. *Vertical navigation* changes the level of abstraction and with it the degree of detail shown in the visualization. Different abstractions can be created by expanding or collapsing clusters to include or exclude their individual nodes from the node-link diagram. *Horizontal navigation* relates to changing which part of the node-link diagram is visible on the screen. Zooming and panning are the typical operations to support horizontal navigation.

Knowledge Gap   A comprehensive exploration of a hierarchical graph usually requires numerous horizontal and vertical navigation steps. However, it is not always easy for users to decide on where they should continue the data exploration, which corresponds to a *target unknown* knowledge gap. Even if users have an idea of the data they want to inspect, it can be difficult to define an appropriate sequence of horizontal and vertical navigational steps to get to the desired target, which is obviously a *path unknown* knowledge gap.

Guidance   To assist users during the visual analysis of hierarchical graphs, it makes sense to offer guidance at two degrees. First, *orienting guidance* can indicate to the user where interesting nodes are located. Second, *directing guidance* can recommend and provide direct access to nodes that are most worth visiting next based on the current exploration situation.

The guidance generation (G*) starts with a search for recommendation candidates. The search takes place in the neighborhood of the data currently being visible to ensure that the recommendations are indeed related to what the user is seeing in the visualization. In fact, the search involves three different neighborhoods. The graph neighborhood relating to distances in the graph, the attribute neighborhood concerning similarities among the attribute values associated with nodes, and the visualization neighborhood as defined by node positions in the graph layout. While the graph neighborhood and the attribute neighborhood are given in [D], the visualization neighborhood takes into account [S] and [I].

Once a set of candidates has been collected, the next step is to select a few of them to be suggested to the user. This can be done by means of a degree of interest (DoI) function, as discussed in Section 5.2.1 in the previous chapter. The DoI function includes several components, such as a priori domain interest given in [D*], user interest from [K], and the distance to the current view as specified in [S]. It further makes sense to model interest degradation for data that have already been visited as stored in [H*].

Finally, the candidates with the highest DoI are presented as navigation recommendations to the user. The visual design should follow a defensive strategy in order to only minimally interfere with regular data exploration.

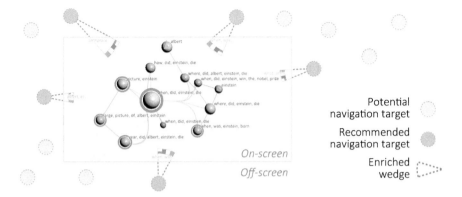

Figure 6.10   Navigation recommendations for graph visualization.

Only when users have difficulties in determining a good next navigation target
on their own should their attention shift to the navigation recommendations.

Figure 6.10 utilizes the *enriched wedges* introduced in Section 4.5.3 in
Chapter 4 to guide users. The wedges themselves serve as visual cues [C*]
that indicate direction and distance of recommended targets for horizontal
navigation. The bars in the wedges visualize the components of the DoI function
to make clear to the user why a target is recommended.

Recommended targets for vertical navigation are by definition not contained
in the currently visualized graph cut, and hence they cannot be pointed
at with an enriched wedge. Therefore, visual cues for vertical navigation
recommendations are attached to anchor nodes whose expansion (or collapse)
would make the recommended target visible. In Figure 6.10, mildly pulsing
rings around anchors suggest that an expand (outward pulsing) or a collapse
(inward pulsing) operation will uncover a target of interest.

Both the enriched wedges and the pulsing rings serve a second purpose.
Each of them is associated with a navigation shortcut to be used as option [O*]
for the data exploration. If the user decides to follow a shortcut by clicking
or taping on it, the cut through the hierarchical graph and the view on the
graph layout are automatically set such that the associated target becomes
visible. This enables the user to get to the target without traversing the path
to it manually. Meanwhile, new navigation recommendations are prepared in
the background based on the new analysis situation. Should further assistance
be necessary, the system can suggest new targets at once.

In summary, this first example of guided visual analysis illustrated how
users can be assisted in making informed navigation decisions when exploring
hierarchical graphs. Next, we will look at an approach for guiding the visual
analysis of large heterogeneous data.

6.2.3   Guiding the Visual Analysis of Heterogeneous Data

The second example investigates how guidance can help physicians exploring heterogeneous biomedical data [Str+12]. The visual analysis is conducted to inform the treatment of newly diagnosed cancer patients.

Data, Analytic and Visual Transformation, Exploration   As illustrated in Figure 6.11, the data involved in the treatment planning are diverse. Physicians have to consult patient-related data from different sources, including anamnesis information, MR, CT, and X-ray images, tissue samples, and lab results. Moreover, general biomedical data have to be considered, including protein and gene expression data, pathways and published articles from medical databases.

The visual analysis of such heterogeneous data naturally involves diverse analytic computations and different visual representations. For example, multi-variate patient data usually need to be filtered before they can be represented in a tabular or parallel coordinates visualization. Gene expressions are typically clustered before being represented in dendrogram heatmaps.

During the data analysis, physicians carry out different exploratory tasks in a step-by-step fashion. For example, they browse patient data to find similar cases, study anamnesis transcripts to understand diagnoses, and inspect related publications, gene expressions, and pathways to learn about latest research results. In the end, the objective is to determine the treatment for a patient.

Knowledge Gap   In the first place, it must be decided which particular part of the heterogeneous data is to be investigated to accomplish the task at hand. Second, suitable analytic and visual transformations must be employed to make the relevant information visible. Third, individual analysis steps must be sequenced properly to gain comprehensive insight into a case. Physicians who are not visualization experts might find these questions difficult to deal with, resulting in *target unknown* and *path unknown* knowledge gaps.

Guidance   The guidance generation (G*) relies on a tailored domain model [D*]. It is constructed by data analysis experts in an extensive modeling phase prior to the visual data analysis. The modeling starts with the individual subsets of the heterogeneous biomedical data. They are shown as larger boxes in Figure 6.11. Each part of the data is then annotated. The annotations indicate which analytical and visual tools can process certain data, and which tasks can be performed by inspecting them.

Finally, links are established between the individual parts of the data to model workflows as sequences of analysis tasks. Each workflow forms the basis for a concrete analysis session that pursues a specific goal. The result of these efforts is a tailored domain model [D*], which will be the basis for guiding the visual analysis.

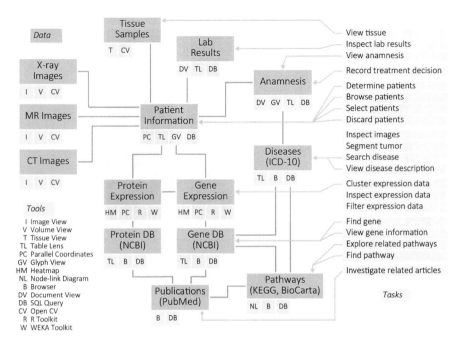

Figure 6.11 Tailored domain model as the basis for user guidance. Adapted from [Str+12].

Two types of guidance support the physicians. *Orienting guidance* is offered to help physicians keep track of the data that have already been explored and those data that are still to be examined. To this end, the current state of the analysis workflow and the data that have already been analyzed are indicated through visual cues [C*]. In addition, *directing guidance* recommends tasks to be accomplished next. The corresponding options [O*] are based on the current analysis path through the domain model. Altogether, the guidance provides physicians with information about where they are, what they have already done, and what they can do now.

Figure 6.12 illustrates the described approach. A series of larger symbols at the bottom of the figure depict the analysis path taken. Possible options for continuing the analysis path are shown as smaller symbols in the bottom-right corner. The highlighted red option hints at the recommended next step. The actual visual analysis is carried out with the visual representations shown in the central part of the figure. The view for the current analysis step faces the user, whereas other views are tilted. Visual links assist the physicians in relating the information being displayed in the different views.

As we have seen, guidance can be as simple as suggesting navigational steps, but also as comprehensive as coordinating many different aspects, including views, methods, and tasks based on domain-specific workflow models.

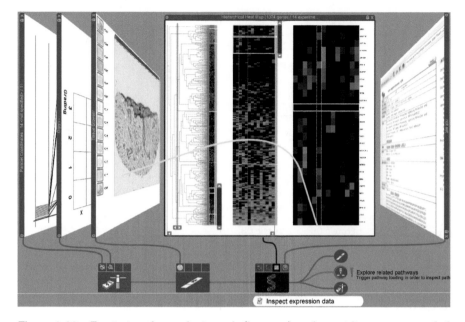

Figure 6.12   Depicting the analysis path (bottom) and providing recommended next steps (bottom-right) can guide the visual analysis of heterogeneous biomedical data. Reprinted from [Str+12].

We can conclude that guidance has the potential to improve the human-computer cooperation in interactive visual data analyses. Guidance has to be unobtrusive to the user, and adaptive to the particular context, as the type of assistance a user requires varies and depends on many factors. Good guidance provided in difficult analysis situations allows users to focus more on a deeper understanding of the interesting phenomena in the data, rather than the means employed for acquiring the understanding.

6.3   PROGRESSIVE VISUAL DATA ANALYSIS

In general, interactive visual data analysis depends on a smoothly running loop of human analytical thinking and system-generated visual representations. Guidance, as described in the previous section, helps us keep the loop alive on the human side. Let us briefly return to van Wijk's original model, but now depicted with a slightly different focus in Figure 6.13. We can see how knowledge [K] is generated *step-by-step* through perception (P), as indicated by dK/dt. Similarly, interactive exploration (E) changes the transformation specification [S] *incrementally*, which is denoted as dS/dt. Guidance aims to keep these human-oriented processes running.

But we can also see that there is one more process participating in the loop. It is the system-oriented transformation (T) of data [D] into images

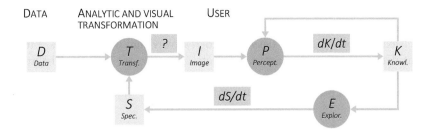

Figure 6.13 Incremental processes highlighted in van Wijk's model of visualization. Adapted from [vWij06].

[I] based on the specification [S]. In van Wijk's model, in our book so far, and in the vast majority of the literature in general, this transformation is considered to be a single, non-incremental step. A corresponding dI/dt is therefore missing. However, transforming large and complex data can require considerable processing time, which can lead to a stuttering analysis loop. This computational challenge of interactive visual data analysis can be addressed by introducing progressive means, which will be discussed in this section.

So, in addition to human aspects, the system architecture is critical for a smooth interactive visual data analysis. Heer and Shneiderman make the following statement on the engineering of efficient visualization infrastructures [HS12]:

"Especially for large datasets, supporting real-time interactivity requires careful attention to system design and poses important research challenges ranging from low-latency architectures to intelligent sampling and aggregation methods."

Heer and Shneiderman, 2012

Bringing in line the cost involved in generating visual representations with the need to present visual feedback immediately is a major technical challenge. A straightforward monolithic implementation of the classic visualization pipeline might not be sufficient in this regard. Any analytical computation, mapping transformation, or graphical operation along the pipeline that fails to deliver results within interactive response time will disrupt fluid interactive analytic work [LH14].

What is needed is an architecture that can cope with complex, time-consuming computations, while still being able to react to interactive user requests and to provide rich visual feedback. One option to address this need is to implement what is called *progressive* visual data analysis. That is, the data transformation (T) is realized in a progressive fashion, which conceptually replaces the question mark in Figure 6.13 with the missing dI/dt.

Next, we introduce the basics of progressive visual data analysis. First, we consider fundamental conceptual aspects. Then we outline a multi-threading architecture for implementing progressive solutions. Finally, we discuss key scenarios where progressive means can enhance the visual data analysis.

6.3.1 Conceptual Considerations

The key idea behind progressive visual analysis is to generate partial visualization results of increasing completeness and correctness [SPG14]. There are two options for breaking down time-consuming computations on large data [Sch+16]:

- **Subdivide computations** into smaller steps.

- **Subdivide data** into smaller chunks.

Performing calculations on smaller chunks of data or in smaller computational steps has three key advantages with respect to:

- responsiveness of the system,

- transparency of the involved calculations, and

- control of the visual analysis.

Taking smaller steps on smaller data chunks allows a system to be responsive to interaction requests of the user. Long-running tasks, which block the system, are avoided and the latency between user request and system response is reduced. Showing partial results in a progressive fashion also facilitates transparency. Particularly analytical methods often come as *black boxes* whose inner workings remain undisclosed to the user. Through progressive approaches, users can observe and thus better understand how analytical calculations converge to a final result. Responsiveness and transparency together improve control of the overall analytic process. Already before the final result is produced, users can steer the analytic process, for example, by prioritizing regions of interest or stopping calculations whose partial results did not yield any fruitful insights.

Modeling the Subdivision of Computations and Data

In the following, we describe how subdivisions of computations and data can be modeled conceptually [Sch+16]. To this end, we build upon the data transformation as introduced in Section 2.3 of Chapter 2. Originally, we described the data transformation as a pipeline of monolithic operators that carry out computations on the entire data and pass on a single result to subsequent operators. Now, for progressive visual data analysis, the notion of operators and transitions between them needs to be refined to accommodate smaller computational steps on chunks of data.

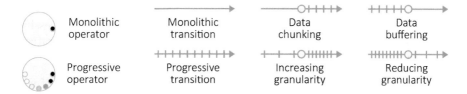

Figure 6.14 Extended notation for operators and transitions for progressive visual data analysis. Adapted from [Sch+16].

Progressive Operators   Progressive operators realize a sub-division of the computational process. To make this clear, the operator notation is extended to include small marks that represent the generated results. As illustrated in Figure 6.14, a monolithic operator generates only a single, complete result. On the other hand, a progressive operator can generate multiple partial results of increasing quality, indicated by increasingly darker shades of gray, until the complete result is produced.

How many results *can theoretically be* produced is operator-dependent. A prerequisite for more than a single complete result is that the underlying algorithms are able to produce results incrementally. Often, classic implementations of known algorithms do not have this property. For example, a sorting algorithm usually does not provide intermediate results. Therefore, it can be necessary to utilize or develop adapted variants that fulfill said requirement.

How many partial results *should sensibly be* produced is user-dependent, because different analytic tasks require different update rates. For example, it may be more helpful to show only those partial results that differ at least by a given Δ from the previously shown one, rather than showing each and every possible partial result.

In any case, each partial result must be a valid input to the subsequent operator that follows down the transformation pipeline. The passing on of data and results is modeled via progressive transitions.

Progressive Transitions   For the classic pipeline, we have not talked about transitions between operators per se, because they trivially transfer the full data. Yet, for progressive visual data analysis, transitions deserve special attention, because they model the subdivision of the data.

In Figure 6.14, a progressive transition can be recognized by a series of marks along its arrow line. These marks denote the data flow to be a stream of data chunks. The distance between the marks indicates the size of the chunks. The smaller the space between consecutive marks, the smaller is the chunk size.

To create data chunks in the first place, a transition for data chunking is required. There are different strategies for chunking the data:

- *Incremental Chunking.* The data are sub-divided into disjoint subsets. Each subset contains a sampled version of the original data, and the union of all subsets represents the entire dataset. Ideally, incremental chunking leads to the key features of the data being visible in the progressive visualization early. The challenge is to find a suitable sampling strategy and an appropriate order for processing the chunks.

- *Semantic Chunking.* The subdivision is based on the semantics of the data and the visualization. For example, for a geographic visualization of movement data, it makes sense to deal separately with geographic boundaries, streets, map tiles, and movement trajectories. Small and relevant semantic chunks should be processed early, whereas larger and less-relevant chunks can be queued in the data stream later.

- *Level-of-Detail Chunking.* The data are sub-divided at different levels of granularity. The result is a hierarchy of chunks, where the root node represents the entire dataset at a high level of abstraction, and the leaves represent smaller subsets with details at finer granularity. Level-of-detail chunking can be done with respect to the data space or the view space, for example, by hierarchical data abstraction or multi-resolution methods, respectively.

The outlined chunking strategies can also be applied in combination. For example, for the progressive visualization of a large graph, it can make sense to create a graph hierarchy where each chunk represents the graph at a different level of abstraction. Per chunk in the hierarchy, nodes and edges can be organized in separate semantic chunks. These can further be sampled into subsets for incremental chunking.

The conceptual counterpart of data chunking is data buffering. It accumulates data chunks or partial results to produce the full data or complete results. For example, visualization images can work with buffering in that they accumulate several partial results of an upstream operator before showing a meaningful or complete result.

To make progressive operators compatible in terms of the granularity of data chunks, it also makes sense to introduce transitions that increase or decrease data granularity along the transformation pipeline. Such granularity changes can be necessary, for example, when one operator requires larger data chunks to create partial results, while a subsequent operator is only able to process smaller pieces of data.

Designing Progressive Visual Data Analysis

With the introduced notations, we can now design a progressive analytical and visual transformation of data into images. Figure 6.15 shows a simple example pipeline [Sch+16]. First, the data are sequenced into chunks, each of them is analytically processed by a monolithic similarity search. The partial results of the search are then mapped to a scatter plot. In order to produce a density map, the partial scatter plots are further sub-divided into smaller chunks. Then,

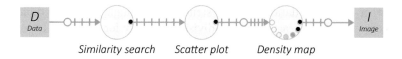

Similarity search   *Scatter plot*   *Density map*

Figure 6.15   Simple example of a progressive transformation pipeline. Adapted from [Sch+16].

the density map operator generates partial results with increasing quality. Finally, the visualization image buffers all partial results into a complete visual representation of the original data.

The design process of such progressive visualization pipelines follows the general design steps for regular visualization solutions as described in Section 2.3 of Chapter 2. Still, there are some peculiarities that deserve special attention.

Meaningful Partial Results   First and foremost, a key requirement of progressive visual data analysis must be kept in mind: The employed progressive means should generate *meaningful* partial results [Sch+16]. That is, the partial results should be interpretable and lead to valid partial insight. What meaningful means concretely depends on the application at hand.

One option to objectively quantify meaningfulness are metrics that measure the quality of partial results. Quality metrics can be calculated with respect to the data or their visual representation. A very simple quality metric is the ratio of the already processed data. The more data have been processed, the higher is the quality. For another example, if a progressive computation involves some statistical error, for example due to the employed data chunking, the error could be used as a quality indicator. Also the differences between successive partial visual representations can indicate result quality. If the difference is rather low, not much new insight can be expected. On the other hand, a significant change between two visual representations could mark the beginning of a progressively unfolding visual pattern.

Progress, error, and differences are valuable indicators that can support users in judging the partial results and steering the analysis. Therefore, designers of progressive visual data analysis solutions should incorporate suitable quality metrics and consider visualizing them along with the actual data. Progress bars should be a mandatory element in the visualization interface.

Design Perspectives   Another aspect of the design is to decide where and how operators and transitions should employ progressive means. These decisions have to take two perspectives into account: the input perspective and the output perspective.

Input Perspective From an input perspective, the data and the involved transformation calculations are relevant. If the data do not fit into the memory, progressive data chunking needs to be considered. If, on the other hand, the data are highly structured, it might be rather difficult to come up with a reasonable sub-division strategy for the data. In terms of computations, if the runtime of an operator is rather long, it makes sense to replace it with a progressive alternative. Yet, if a time-consuming operator produces merely optional decorations, it may be more sensible to leave the operator as is, because it is acceptable to wait for the operator to finish in the few cases this is indeed needed.

Output Perspective From the output perspective, the tasks and interests of the analyst are relevant. The designer needs to know whether the analysis follows an overview-first strategy or a detail-first strategy.

With an *overview-first strategy*, the goal is to show the complete data as quickly as possible, where a lower degree of detail is initially acceptable. Following this strategy means progressively presenting data abstractions with increasing granularity. First, a coarse overview is displayed based on which the user can start interpreting the data. Looking into regions of interest and studying specific details becomes possible as more and finer-grained information is being transmitted progressively.

In contrast to that, the *detail-first strategy* prioritizes highly detailed representations, where it is acceptable to see only subsets of the data in the beginning. The detail-first strategy progressively visualizes chunks of the original data, rather than data abstractions. Once the first chunk has been transmitted, a detailed view of a specific piece of data is presented to the user. During the progression, more and more pieces will be added until the visualization shows the complete data. Note, however, that the visibility of details is gradually diminished the more data chunks are being presented.

In summary, this section described the basic building blocks of progressive visual data analysis and the design requirements and perspectives to be considered. Yet, these conceptual aspects are only one part of progressive visual data analysis. Another important practical question is how to implement the underlying progressive software. This question will be addressed in the following.

6.3.2 Multi-threading Architecture

Progressive visual data analysis software must be able to generate and show partial results while still being responsive to user requests. Therefore, it is practical to follow a *multi-threading* architecture [Pir+09]. A schematic outline of such an architecture is given in Figure 6.16. It consists of the data, the specification, and the image artifacts, which we already know from earlier figures. Additionally, there are two central computing components: the *control thread* and the *processing threads*.

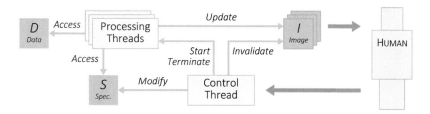

Figure 6.16 A multi-threading architecture for progressive visual data analysis.

Control Thread   The control thread is in charge of receiving user input and coordinating the other components. As these tasks are usually easy to accomplish, the overall system stays responsive at all times. The typical workflow of the control thread is as follows. When an interaction request arrives, the control thread modifies the specification and invalidates the visualization. Additionally, any processing threads that operate with the old specification are terminated early because they would no longer generate valid, but obsolete results. Finally, new processing threads are started to generate new valid visual output.

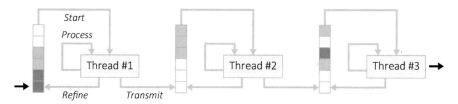

Figure 6.17 Illustration of asynchronous processing threads operating on data chunks stored in priority queues.

Processing Threads   The processing threads implement the progressive operators introduced before. As illustrated in Figure 6.17, these threads have a priority queue of data chunks to be processed. When incoming data chunks are appended to the queue, the processing thread starts working. It removes a data chunk from the queue and processes it in an internal loop until a given break condition is reached. This condition can be based on the processing time spent or on the quality of the partial result. If the partial result is of sufficient quality, it can be transmitted to the subsequent thread's priority queue. If the result is not yet good enough, it is re-appended to the thread's own queue for further refinement.

The different shades of gray in Figure 6.17 indicate how close the queued chunks are to the final result. These shades correspond directly to the marks in the progressive operators, where white stands for unprocessed chunks, gray for intermediate results, and black for the final result.

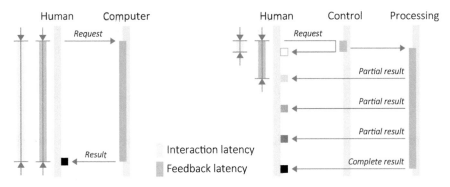

(a) Regular single-thread solution.   (b) Progressive multi-thread solution.

Figure 6.18   Comparison of single-thread and multi-thread solutions.

As soon as meaningful results reach the end of the pipeline, the visualization gets updated. Here it makes sense to organize the visual output into multiple layers according to the different data chunking strategies (semantic layers, incremental layers, or level-of-detail layers). Using multiple layers enables the architecture to provide rich and scalable visual feedback, and to avoid redundant computations by reusing cached results that remain valid after a user interaction.

If a processing thread's queue is empty, the thread stops working. A processing thread can also be terminated early at any time by the control thread, either by emptying its queue or by dismissing thread and queue altogether.

A key advantage of the described multi-threading architecture is the reduction of latency by generating partial results. This is made clear in Figure 6.18. The classic monolithic approach is depicted in Figure 6.18a. As indicated by the colored vertical bars, the human has to wait for a while until a single complete response is generated. While the system is busy producing the complete response, the system is unresponsive and no feedback about the ongoing computations is delivered to the user.

In Figure 6.18b, we can see how the progressive multi-threading architecture reduces the latency. The interaction latency is much shorter and the system remains responsive at all times, because the control thread handles only brief coordination tasks. The processing thread works on the data in the background and provides meaningful partial results as soon as they are ready, which reduces the feedback latency.

Up to this point, we have discussed how progressive visual data analysis can be modeled, designed, and implemented. The next section will illustrate how progressive means can be put to use to support a smooth interactive visual data analysis.

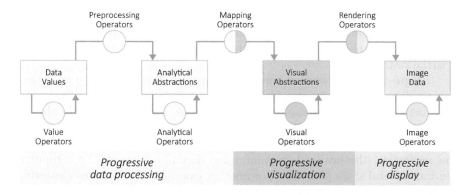

Figure 6.19 The three typical scenarios of progressive visual data analysis: progressive data processing, progressive visualization, and progressive display.

6.3.3 Scenarios

Operating progressively is a general approach and any phase of the analytical and visual data transformation can benefit from it. Depending on which phases primarily utilize progressive means, three typical scenarios can be identified: progressive data processing, progressive visualization, and progressive display.

Figure 6.19 illustrates the three scenarios and their relation to the phases of the visualization pipeline. Progressive data processing subsumes all data chunking and computational operators that work on data values and analytical abstractions. Progressive visualization covers chunking and operators that are used to create and manipulate visual abstractions. Finally, progressive display deals with chunking and operators at the level of image data. Next, we discuss the three scenarios of progressive visual data analysis in more detail.

Progressive Data Processing

Given the fact that progressive data processing takes place in the beginning of the analytical and visual transformation, it is primarily concerned with the input perspective, that is, the properties of the data and the calculations. Progressive data processing is typically applied to cope with two challenging situations. First, if the data are too large to be processed in a timely manner all at once, progressive data processing can provide early feedback that allows users to see in which direction the calculations are going, and if necessary, to steer the calculations toward more desirable results. Second, if the data processing involves algorithmic black boxes or the data are unknown, looking at progressively generated partial results can help users better understand both the data and the operations performed on them.

Progressive value operators, preprocessing operators, and analytical operators aim at an efficient and effective handling of the data by sub-dividing the involved computations. An example of a progressive value operator is adaptive sampling. Adaptive sampling delivers data samples successively and allows users to adapt the sampling strategy on the fly. This way, the representativeness of data being analyzed can be improved based on insight gained from early partial results.

Progressive preprocessing operators transform data values into analytical abstractions in a step-by-step fashion. Chapter 5 discussed several fundamental methods for creating analytical abstractions. Many of them can also be employed in a progressive setting. For example, if the data to be analyzed do not fit into memory, a principal component analysis can be done progressively. In particular, the involved singular value decomposition can be computed in incremental steps [Sar+02]. For another example, if a k-means clustering is part of the analysis pipeline, but users are unfamiliar with the data and hence do not know what a suitable k is, it makes sense to employ iterative clustering [Kim+17]. It allows users to adjust the number of clusters or the position of cluster representatives on the fly while already examining the partial clusters being produced.

Progressive analytical operators are concerned with computations on the derived analytical abstractions. This can involve structuring, organizing, and prioritizing the abstractions. For example, data abstractions can be organized in hierarchical data structures. When traversing such data structures, it is sensible to prioritize data chunks representing larger data deviations early. As we know from Section 5.3.2 of Chapter 5, larger deviations tend to add more information to the visualization than chunks with only smaller changes.

The progressive operators described so far sub-divide the computations on data. Yet, with very large data, it is hardly possible to deliver results in interactive frame rates. Therefore, a second key concern of progressive data processing is to sub-divide the data into chunks using a suitable strategy (incremental chunking, semantic chunking, or level-of-detail chunking). The following example will briefly illustrate the positive effect that chunking can have on the visual data analysis.

The example concentrates on the first part of the progressive pipeline presented earlier in Figure 6.15. The dataset to be processed comes from the Fatality Analysis Reporting System of the US National Highway Traffic Safety Administration and contains information about more than 370,000 car crashes between 2001 and 2009. When analyzing the crashes with respect to certain criteria, a similarity search must be performed on the data. Yet, searching the whole dataset is computationally expensive. Therefore, the search operates on chunks of 5,000 crashes instead. This ensures reasonably quick feedback for interactive adjustments [Sch+16].

Figure 6.20a illustrates how the chunked search progressively fills the display with crashes matching the search criteria. During the process, users can update the criteria as needed. Moreover, they can define a region of interest in order to prioritize the data chunks associated with that region. Such a prioritized progression is illustrated in Figure 6.20b. The effectiveness of the visual data analysis is increased because interesting data can be inspected early.

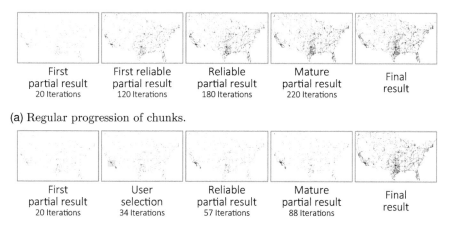

(a) Regular progression of chunks.

(b) Prioritized progression of chunks.

Figure 6.20   Visualization of progressively processed data chunks of car crashes from a database with more than 370,000 entries. Courtesy of Marco Angelini.

. This section provided just a glimpse on what is possible in terms of progressive data processing. Next, we will discuss the scenario of progressive visualization.

Progressive Visualization

Progressive visualization is primarily concerned with the graphically oriented transformation steps that map data to visual abstractions and render them to visualization images. Progressive visualization is typically applied when the amount of data or visual abstractions is excessively large, or if the involved mapping and rendering operations are computationally expensive. In such cases, it is usually not feasible to process everything in a single step. Even if it was, the visualization could be too dense for the user to interpret it with ease.

Progressive visualizations utilize chunking and progressive computations to build up visual representations step-by-step. The individual steps are computationally easier to deal with and the user has a chance to observe the construction process and hence can gain a better understanding of the visualization design and the data.

Picking up on the output strategies *overview-first* and *detail-first* discussed earlier, we will next present two examples that illustrate progressive visualization.

Progressive Force-directed Graph Layout   As we know from Section 3.5 of Chapter 3, node-link diagrams are a common representation for graphs. Nodes are basically represented as dots and edges are shown as links between

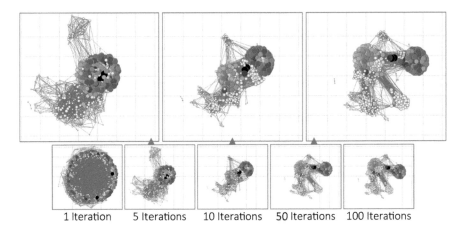

1 Iteration   5 Iterations   10 Iterations   50 Iterations   100 Iterations

Figure 6.21   Progressive force-directed layout of a social network with 747 nodes and 60,050 edges.

dots. These visual abstractions are easy to create. Yet, a key challenge is to determine a suitable layout of the dots. This requires sophisticated graph layout algorithms, which are usually computationally expensive.

For our example, we utilize the force-directed layout approach, which is known to generate visually pleasing results and also to have a high runtime complexity due to the involved simulation of the forces. Yet, as the simulation is iterative by nature, it perfectly fits the scenario of progressive visualization. Following the overview-first strategy, the progression starts with a rough layout that contains the entire graph. With each iteration of the simulation, the layout is refined until a high-quality output is obtained. High quality means that the simulated forces are in equilibrium.

Figure 6.21 shows a series of snapshots of a progressively unfolding node-link diagram of a social network with 747 nodes and 60,050 edges. One can nicely observe how the layout algorithm separates the key structures in the data between the 5th and 50th iteration.

Progressive Network Mapping   Our second example will illustrate the detail-first strategy applied to a climate network with 6,816 nodes and 232,940 edges. Climate researchers use such networks to study climate change on Earth. The nodes of the network span a grid across the globe, so there is no need to compute a graph layout. Yet, the number of visual abstractions is relatively high.

For our example, we visualize the climate network on a 3D globe. Each node is represented as a 3D sphere and each edge is shown as a 3D curved link. Creating these 3D visual abstractions poses significant stress on the graphics hardware. Therefore, it makes sense to process the data in chunks. First, semantic chunking is applied: The nodes form one chunk and the edges form

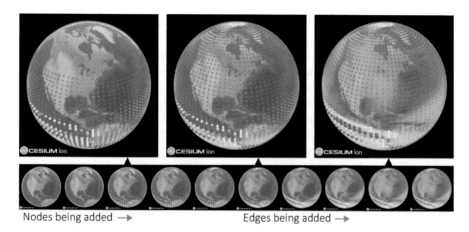

Figure 6.22   Progressive visualization of a climate network with about 6,816 nodes and 232,940 edges.

another chunk. Because there are still very many nodes and edges per chunk, an additional sub-division by incremental chunking takes place. Eventually, we obtain seven chunks of nodes and 233 chunks of edges, each holding a thousand data elements. The progressive visualization will process these chunks one after the other.

The result is depicted in Figure 6.22. According to the detail-first strategy, the progression starts with full details, but only for a part of the data. The node chunks come first followed by the chunks of edges. The first pieces of data are presented to the user after about five seconds. Without progressive chunking, the user would see a blank screen for about 40 seconds.

The above examples illustrate quite well how progressive visualizations gradually improve the visual output before they ultimately end in a refined and complete final result. The gradual build-up of visual representations contributes to a better understanding of the visualization and the data.

Progressive Display

Last but not least, the progressive display of visual representations is a scenario where progressive techniques are employed to support the visual data analysis. This scenario brings us closer to methods for progressive encoding, transmission, and decoding of image data, which have a long history in computer graphics research.

Progressive display techniques are particularly interesting in environments with limited connection bandwidth and in environments with heterogeneous output devices, such as the smart meeting room described in Section 6.1. Based on a well-prepared stream of image data, progressive techniques can ensure that important image characteristics are transmitted first, whereas additional details are progressively fetched to complete the view. Moreover, individual output devices can acquire exactly the amount of image data they need for their particular size and resolution. Smaller low-resolution devices can stop the transmission early, whereas larger high-resolution displays extract the full pixel information from the data stream [Ros+11].

Figure 6.23 illustrates such a device-dependent progressive transmission of a treemap visualization. The treemap is to be displayed on heterogeneous displays in a smart meeting room. Note that there is only one static visualization image. The progressive image viewer running on each display will extract and transmit exactly the pixels that match the device resolution.

Progressive display is not limited to device-dependent adaptations. It is also possible to display visualization images progressively based on user interests [Rad+15]. For example, when a discussion in a smart meeting room shifts from a global perspective to local areas, the progressive display can provide further image data allowing users to see the areas of interest at higher resolution.

Figure 6.24 shows an example with a color-coded map of the world. Once the user has marked a region of interest, additional image data are progressively transmitted to enhance the visualization. It will then be shown either with an *overview+detail* approach or as a *focus+context* display. Both overview+detail and focus+context have already been introduced as fundamental visual arrangements in Section 3.1.2 of Chapter 3. Note again that the progressive display operates only on the visualization image. There is no need to reprocess the entire dataset in order to obtain interest-dependent views of it. Whenever user interests change, new regions of interest can be marked and additional image data is progressively transferred accordingly.

Technically, a key requirement is that the visualization images be encoded in such a way that progressive image transmission and refinement are possible. Progressive JPEG, for example, can present an overview of the image and then refine the image successively. The JPEG2000 standard adds support for the

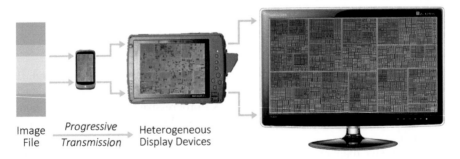

Image File   *Progressive Transmission*   Heterogeneous Display Devices

Figure 6.23   Device-dependent progressive transmission of a treemap visualization image. Adapted from [Ros+11].

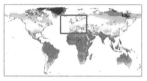  

(a) Global view.　　　(b) Overview+detail view.　(c) Focus+context view.

Figure 6.24　Progressive display for a dynamically defined region of interest. Courtesy of Axel Radloff-Delosea.

definition and prioritized transmission of regions of interest. The two examples of the device-dependent treemap visualization and interest-dependent map visualization are based on JPEG2000-encoded image data [RS09].

With the three scenarios of progressive data processing, progressive visualization, and finally progressive display, this section on progressive techniques for interactive visual data analysis is complete. Overall, allowing fluid interaction and providing timely feedback are important aspects for visual data analysis. Studies provide first evidence of the positive impact of progressive techniques [Zgr+17]. Users can generate more insights per minute and are more agile in their interactions when working with progressive visualizations. Meaningful partial results can be employed to pre-process information, extract insights early, or decide quickly that an alternative visualization could be more useful for the task at hand. All these benefits suggest that progressive visual data analysis is an important approach when exploring large datasets.

6.4　SUMMARY

Research on interactive visual data analysis is advancing on various fronts. In this chapter, we discussed a selection of three advanced topics: advanced visualization in multi-display environments, guidance as advanced support for interactive analysis, and progressive techniques for advanced processing of analytical and visual data abstractions. It is in the nature of such advanced topics that they have typically not yet been fully investigated in research. There are still open issues to be discussed and remaining challenges to be addressed.

Recent works on multi-device visualization further study the seamless integration of multiple devices to form coherent visual analysis environments [Kis+17; LKD19; Hor+19]. Interesting questions are how to best utilize displays of different form factor and their different interaction modalities, how to support collaborative exploration in dynamic multi-devices ensembles, and how to technically cope with a heterogeneous zoo of platforms and infrastructures.

With respect to guidance for interactive visual data analysis, more research is needed on designing and implementing actual guidance solutions [Cen+17]. This involves questions such as: how can knowledge gaps and the required degree of guidance be inferred, when is the right moment to provide guidance, what are suitable methods to convey guidance to the user, and how can the success or failure of guidance be evaluated?

As indicated in the previous section, progressive techniques bear much potential to enhance interactive visual data analysis. Yet, research on this topic is still young and there are many questions to be investigated [Fek+19]. Already the conceptual definition and fundamental theories behind progressive techniques are subject to discussions. Moreover, the benefits and also threats of progressive techniques need to be studied in more depth from a human perspective.

As a conclusion, we can state that the advanced methods discussed in this chapter illustrate quite nicely how data analysis can go beyond plain visualization, interaction, and computation. Yet, it is also clear that these advanced methods are not yet mature and that it will certainly need some time before they appear in regular data analysis solutions for day-to-day use.

FURTHER READING

Multi-display Visualization: [CD04] • [Chu+15] • [Rad+15] • [EST19]

Guidance: [Hor99] • [Cen+17] • [Col+18]

Progressive Analysis: [Fis+12] • [Müh+14] • [SPG14] • [BEF17] • [Ang+18]

Summary

CONTENTS

T HIS BOOK on interactive visual data analysis comes to an end. We worked through six chapters on visual, interactive, and analytical methods for making sense of data. In this final chapter, we will briefly recapitulate the key points of the book and provide you with ideas on how to continue once you put this book down.

7.1 WHAT'S BEEN DISCUSSED

In **Chapter 2**, we introduced fundamental aspects. In the first place, there are the three key criteria of interactive visual data analysis: expressiveness, effectiveness, and efficiency. Expressiveness tells us that visual representations should show us exactly what is in the data and that interaction should allow us to do exactly what is needed to understand the data. Effectiveness calls for the consideration of human factors. Efficiency is satisfied when the costs and the benefits of the interactive visual analysis are balanced.

Further, we discussed the data, the tasks, and the context as three key influencing factors of interactive visual data analysis. While the data properties characterize *what* is to be analyzed, the task aspect is concerned with *why* data are analyzed. The context, describes by *whom* and *where* the data analysis is carried out.

Finally, we studied models of the processes that take place at different stages of interactive visual data analysis. We considered the design process as a cascade of individual design steps, the process of transforming the data into visual representations as a pipeline of operators, and the knowledge generation process as analysis loops that produce new findings and insight.

Chapter 3 was dedicated to visualization. We started with basic methods for encoding data visually and presenting them to the user in a meaningful way. Visual variables such as position, length, or color hue were introduced as the fundamental means to convey information graphically. In fact, visual variables modify the appearance of graphical marks, which allows us to perceive visual differences and interpret the depicted information.

Visual variables and marks are the basic building blocks of visualization techniques. We introduced many different techniques for different data classes. For multivariate data, we described table-based, combinded bivariate, polyline-based, glyph-based, pixel-based, and nested visualizations. They can be considered the basic tools of the trade. Then we moved on to visualization techniques for temporal data and geo-spatial data, and combinations thereof. We characterized time and geographic space as special dimensions and presented various dedicated visualizations taking into account the specifics of time and space.

Last but not least, our interest concerned the visualization of graph data. Graphs describe not only data elements per se, but also relations between data elements. Basic graph visualization techniques show these relations as links between elements or via clever arrangements of elements. We also learned how multi-faceted graphs can be visualized by combining ideas from different visualization techniques.

Following the chapter on visualization was **Chapter 4** on interacting with visualizations. We begun the chapter with a description of various scenarios where it is useful or even necessary to let the user interact. What "interacting" actually means was abstractly modeled as a cycle of actions. In order to allow this cycle to run smoothly, several requirements and guidelines need to be considered.

In terms of interaction for visual data analysis, we built upon basic operations and discussed fundamental selection and accentuation techniques. To address the analysis of large data, a whole section was dedicated to navigating zoomable visualizations. Zooming is a fundamental technique to satisfy the need for an overview of the data and also for finer details of selected parts of the data. Interactive lenses were introduced as a versatile lightweight means for locally adapting visual representations of data. The potential of naturally inspired interaction was illustrated for the important task of visual comparison.

Leaving classic desktop interaction behind, we moved on to modern interaction technologies and learned how they can enhance visual data analysis. Touch interaction makes working with visual representations of data truly direct. Tangible views provide an extended interaction vocabulary making it possible to analyze data in new and interesting ways. Finally, with proxemic interaction it is possible to explore more and larger data on high-resolution display walls.

Chapter 5 taught us that large and complex data also require analytical support through automatic computation. The primary goal was to reduce the complexity of the data and their visual representations. Density-based representations and bundling were introduced as methods to reduce complexity on the visual side.

Reducing the complexity of the data can be done in various ways. We considered degree-of-interest approaches and feature-based methods as a means to focus on relevant data. Another option is to abstract the data via sampling and aggregation. Data abstraction can be done repeatedly to generate multi-scale data abstractions.

Automatic computations can also serve to group data elements. We introduced two fundamental approaches: classification and clustering. Classification is a means to sub-divide the data space, and clustering groups data elements according to a certain similarity measure. How complex multivariate dynamic graphs can be clustered was discussed in a separate section.

Finally, we described principal component analysis as a tool to address the challenge of analyzing data with very many variables. The key idea here was to project the original high-dimensional data space to a lower-dimensional space.

In the end, **Chapter 6** provided an outlook on advanced topics in interactive visual data analysis research. Following the three-fold structure of the previous chapters, we discussed advanced visualization, advanced interaction, and advanced automatic computation.

Presenting data in smart multi-display environments can be considered an advanced form of visualization. This makes it possible for multiple users to analyze data collaboratively. We also saw that several mechanisms must be integrated into the environment to support the data analysis.

User guidance was introduced as an advanced means to support interaction when users have difficulties in making analytical progress. We elaborated on a characterization of guidance and a conceptual model.

Our last advanced concept considered the use of progressive methods to enhance automatic computation. We made a step from otherwise monolithic operators that generate a single complete result to progressive operators that generate several partial results with increasing quality.

7.2   HOW TO CONTINUE

Now that this book is over, we want to point you to ideas of how you can continue with the topic of interactive visual data analysis. We think of four key avenues for your next activities: learn, apply, create, and advance.

Learn   A first activity could be to go on and study interactive visual data analysis further. Certainly, this book offered a broad overview, but it could not discuss all aspects in full detail. Therefore, each chapter ended with a list of references recommended for further reading. The lists contain excellent books that cover topics we could only touch here briefly. We also included research articles with in-depth investigations of selected topics of interest.

We particularly recommend learning more about human factors. Perception and cognition are key to interactive visual data analysis. Learning more about them will definitely pay off. Another direction for further studies are the fields of volume visualization and flow visualization. They are concerned with potentially time-varying 3D volumetric data and vector fields. Such data are particularly relevant in medicine and engineering.

Apply If you are a practitioner and have not yet used interactive visual methods, your next step might be to apply some of the methods and techniques discussed in this book to your analytic domain problems. Interactive visual data analysis can be beneficial in all application domains where data are available, which means in virtually any domain.

In the financial sector, visual representations can help us understand transactions or detect fraud. In urban planning, they can help us build models of human mobility. Climate researchers rely on visual representation to investigate the impact of climate change and to communicate their findings to the general public. In humanities, interactive visual data analysis can be applied to reveal hidden relationships in large text corpora. Sociologists can be supported in predicting election outcomes based on visual representations of stance in social media.

Even the classic spreadsheet applications already offer basic visualization capabilities. Tableau, QlikView, and Plotly are flagships among the software packages for interactive visual data analysis in general. Many other dedicated tools exist for specific application problems. For example, Gephi, Cytoscape, and Tulip are dedicated to the visual analysis of graphs. KNIME is a software that focuses on the analytics part.

As you can see, there are many options to apply the knowledge gained in this book. It is up to you to introduce or strengthen interactive visual data analysis in your particular application domain.

Create If you are intrigued by the examples presented in this book or if the existing tools do not meet the requirements of your application problem, go ahead and create your own interactive visual analysis techniques. Depending on your level of expertise, you can extend the software mentioned before or implement a tailored visualization solution from scratch.

Readers with programming skills can resort to open source software libraries. For example, D3.js is the quasi-standard for web-based visualization. The follow-up project Observable even allows multiple people to develop visualization in a team effort. Many other visualization toolkits and libraries exist, for example, the Visualization Toolkit (VTK) and Processing for general-purpose applications, GraphStream and sigma.js specifically for visualizing graphs, or mapbox and CesiumJS for geo-visualization.

Writing software for visualizing larger data typically requires advanced programming skills. Analytic computations should be distributed on multiple computing threads and the visualization should utilize the enormous power of modern graphics cards. Working on such a low level with APIs such as OpenMP, OpenCL, OpenGL, WebGL, DirectX, or Vulkan might be more difficult, but it will allow you to visualize millions of data elements at interactive frame rates.

Advance   Interactive visual data analysis is a relatively young field of research. So your next step could be to advance the field. There are still many unsolved problems to be tackled and new discoveries to be made. Visual data analysis has found its way into smart multi-display environments. And it will be used in many different environments, including large display walls, industrial settings, or mobile smart watches. How visualization, interaction, and automatic computation can be designed to scale with these heterogeneous environments is an open question.

Dealing with ever-increasing data remains a major challenge. Progressive data analysis has been discussed as a suitable means to address this challenge. But we are still lacking the concepts and models to broadly apply progression beyond the simple examples illustrated in this book.

Increasing data size and increasing data complexity also lead to more complex tools. How can we support users in handling these tools? Providing user guidance is one option. However, when is the right moment to provide guidance and what is the appropriate amount of guidance with respect to the situation at hand?

When we think about guidance for users, we could also consider guidance for designers and developers. How should a visualization be designed given certain data characteristics and analysis tasks? Usually, coming up with a good answer requires a good amount of experience. Wouldn't it be great if we had actionable guidelines to come to our aid? Yes, it would, but we are not there yet.

We could go on listing research questions for pages. As a matter of fact, there is still much to be investigated. The major scientific conferences on interactive visual data analysis testify to this fact. VIS, EuroVis, and PacificVis publish innovative research results annually. Journals such as *IEEE Transactions on Visualization and Computer Graphics, the Computer Graphics Forum, and Information Visualization* are great for submitting original articles on visual analytics research. In related fields, such as human-computer interaction, cognitive psychology, data mining, or machine learing, it is also possible to publish—and to learn.

This concludes our broad overview of concepts, models, methods, and techniques, basic and advanced ones, for interactive visual data analysis. Given what you have learned from reading this book, you have everything now to make an informed choice on where to venture next in this exciting field. Among all the possible directions of learning, applying, creating, and advancing interactive visual data analysis, with one being as exciting as the other, you cannot choose wrongly.

Bibliography

[AA06]   Andrienko, N. and Andrienko, G. *Exploratory Analysis of Spatial and Temporal Data – A Systematic Approach.* Springer, 2006. DOI: 10.1007/3-540-31190-4 (cited on pages 31, 127).

[Abe+14]   Abello, J., Hadlak, S., Schumann, H., and Schulz, H.-J. "A Modular Degree-of-Interest Specification for the Visual Analysis of Large Dynamic Networks". In: *IEEE Transactions on Visualization and Computer Graphics* 20.3 (2014), pp. 337–350. DOI: 10.1109/TVCG.2013.109 (cited on pages 215, 217–219, 265, 346).

[Aig+05]   Aigner, W., Miksch, S., Thurnher, B., and Biffl, S. "PlanningLines: Novel Glyphs for Representing Temporal Uncertainties and their Evaluation". In: *Proceedings of the International Conference Information Visualisation (IV)*. IEEE Computer Society, 2005, pp. 457–463. DOI: 10.1109/IV.2005.97 (cited on pages 93, 94).

[Aig+11]   Aigner, W., Miksch, S., Schumann, H., and Tominski, C. *Visualization of Time-Oriented Data.* Springer, 2011. DOI: 10.1007/978-0-85729-079-3 (cited on pages 87, 127, 168, 261, 262, 346).

[All83]   Allen, J. F. "Maintaining Knowledge about Temporal Intervals". In: *Communications of the ACM* 26.11 (1983), pp. 832–843. DOI: 10.1145/182.358434 (cited on page 83).

[AMA07]   Archambault, D. W., Munzner, T., and Auber, D. "TopoLayout: Multilevel Graph Layout by Topological Features". In: *IEEE Transactions on Visualization and Computer Graphics* 13.2 (2007), pp. 305–317. DOI: 10.1109/TVCG.2007.46 (cited on page 118).

[AN13]   Andrews, C. and North, C. "The Impact of Physical Navigation on Spatial Organization for Sensemaking". In: *IEEE Transactions on Visualization and Computer Graphics* 19.12 (2013), pp. 2207–2216. DOI: 10.1109/TVCG.2013.205 (cited on page 204).

[And+10]   Andrienko, G., Andrienko, N., Demšar, U., Dransch, D., Dykes, J., Fabrikant, S. I., Jern, M., Kraak, M.-J., Schumann, H., and Tominski, C. "Space, Time and Visual Analytics". In: *International Journal of Geographical Information Science* 24.10 (2010), pp. 1577–1600. DOI: 10.1080/13658816.2010.508043 (cited on page 96).

[And+13]   Andrienko, G., Andrienko, N., Bak, P., Keim, D., and Wrobel, S. *Visual Analytics of Movement*. Springer, 2013. DOI: 10.1007/978-3-642-37583-5 (cited on pages 127, 225).

[Ang+18]   Angelini, M., Santucci, G., Schumann, H., and Schulz, H.-J. "A Review and Characterization of Progressive Visual Analytics". In: *Informatics* 5.3 (2018), p. 31. DOI: 10.3390/informatics5030031 (cited on page 304).

[AS05]   Amar, R. A. and Stasko, J. T. "Knowledge Precepts for Design and Evaluation of Information Visualizations". In: *IEEE Transactions on Visualization and Computer Graphics* 11.4 (2005), pp. 432–442. DOI: 10.1109/TVCG.2005.63 (cited on page 18).

[AS94]   Ahlberg, C. and Shneiderman, B. "Visual Information Seeking: Tight Coupling of Dynamic Query Filters with Starfield Displays". In: *Proceedings of the SIGCHI Conference Human Factors in Computing Systems (CHI)*. ACM Press, 1994, pp. 313–317. DOI: 10.1145/191666.191775 (cited on page 149).

[Asi85]   Asimov, D. "The Grand Tour: A Tool for Viewing Multidimensional Data". In: *SIAM Journal on Scientific and Statistical Computing* 6.1 (1985), pp. 128–143. DOI: 10.1137/0906011 (cited on page 69).

[AWP97]   Andrews, K., Wolte, J., and Pichler, M. "Information Pyramids: A New Approach to Visualising Large Hierarchies". In: *Proceedings of the IEEE Visualization Conference (Vis)*. Late Breaking Hot Topics. IEEE Computer Society, 1997, pp. 49–52 (cited on page 117).

[Bac+17]   Bach, B., Dragicevic, P., Archambault, D. W., Hurter, C., and Carpendale, S. "A Descriptive Framework for Temporal Data Visualizations Based on Generalized Space-Time Cubes". In: *Computer Graphics Forum* 36.6 (2017), pp. 36–61. DOI: 10.1111/cgf.12804 (cited on page 127).

[Bat+99]   Battista, G. D., Eades, P., Tamassia, R., and Tollis, I. G. *Graph Drawing: Algorithms for the Visualization of Graphs*. 1st edition. Prentice Hall, 1999 (cited on page 113).

[BC87]   Becker, R. A. and Cleveland, W. S. "Brushing Scatterplots". In: *Technometrics* 29.2 (1987), pp. 127–142. DOI: 10.2307/1269768 (cited on pages 149, 157).

[BCK08]   Boriah, S., Chandola, V., and Kumar, V. "Similarity Measures for Categorical Data: A Comparative Evaluation". In: *Proceedings of the SIAM International Conference on Data Mining (SDM)*. Society for Industrial and Applied Mathematics, 2008, pp. 243–254. DOI: 10.1137/1.9781611972788.22 (cited on page 191).

[Bed11] Bederson, B. B. "The Promise of Zoomable User Interfaces". In: *Behaviour & Information Technology* 30.6 (2011), pp. 853–866. DOI: 10.1080/0144929X.2011.586724 (cited on pages 159, 206).

[BEF17] Badam, S. K., Elmqvist, N., and Fekete, J.-D. "Steering the Craft: UI Elements and Visualizations for Supporting Progressive Visual Analytics". In: *Computer Graphics Forum* 36.3 (2017), pp. 491–502. DOI: 10.1111/cgf.13205 (cited on page 304).

[Beh+16] Behrisch, M., Bach, B., Riche, N. H., Schreck, T., and Fekete, J. "Matrix Reordering Methods for Table and Network Visualization". In: *Computer Graphics Forum* 35.3 (2016), pp. 693–716. DOI: 10.1111/cgf.12935 (cited on page 115).

[Bel61] Bellman, R. E. *Adaptive Control Processes: A Guided Tour*. Princeton University Press, 1961. DOI: 10.1002/nav.3800080314 (cited on page 258).

[Ber67] Bertin, J. *Sémiologie Graphique*. Gauthier-Villars, 1967 (cited on page 54).

[Ber81] Bertin, J. *Graphics and Graphic Information-Processing*. de Gruyter, 1981 (cited on pages 3, 131).

[Ber83] Bertin, J. *Semiology of Graphics (W. J. Berg, trans)*. University of Wisconsin Press, 1983 (cited on page 54).

[BH94] Bederson, B. B. and Hollan, J. D. "Pad++: A Zooming Graphical Interface for Exploring Alternate Interface Physics". In: *Proceedings of the ACM Symposium on User Interface Software and Technology (UIST)*. ACM Press, 1994, pp. 17–26. DOI: 10.1145/192426.192435 (cited on page 206).

[BHvW00] Bruls, M., Huizing, K., and van Wijk, J. J. "Squarified Treemaps". In: *Proceedings of the Joint Eurographics - IEEE TCVG Symposium on Visualization (VisSym)*. Springer, 2000, pp. 33–42. DOI: 10.1007/978-3-7091-6783-0_4 (cited on page 117).

[Bie+93] Bier, E. A., Stone, M. C., Pier, K., Buxton, W., and DeRose, T. D. "Toolglass and Magic Lenses: the See-Through Interface". In: *Proceedings of the Annual Conference on Computer Graphics and Interactive Techniques (SIGGRAPH)*. ACM Press, 1993, pp. 73–80. DOI: 10.1145/166117.166126 (cited on page 206).

[BLS99] Brandstädt, A., Le, V. B., and Spinrad, J. P. *Graph Classes: A Survey*. SIAM, 1999. DOI: 10.1137/1.9780898719796 (cited on page 112).

[BMG10] Ballendat, T., Marquardt, N., and Greenberg, S. "Proxemic Interaction: Designing for a Proximity and Orientation-Aware Environment". In: *Proceedings of the International Conference on Interactive Tabletops and Surfaces (ITS)*. ACM Press, 2010, pp. 121–130. DOI: 10.1145/1936652.1936676 (cited on page 203).

[Bor+13]   Borgo, R., Kehrer, J., Chung, D. H. S., Maguire, E., Laramee, R. S., Hauser, H., Ward, M., and Chen, M. "Glyph-based Visualization: Foundations, Design Guidelines, Techniques and Applications". In: *Eurographics 2013 - State of the Art Reports.* Eurographics Association, 2013, pp. 39–63. DOI: 10 . 2312 / conf / EG2013/stars/039-063 (cited on page 73).

[BR03]   Baudisch, P. and Rosenholtz, R. "Halo: A Technique for Visualizing Off-Screen Objects". In: *Proceedings of the SIGCHI Conference Human Factors in Computing Systems (CHI).* ACM Press, 2003, pp. 481–488. DOI: 10 . 1145 / 642611 . 642695 (cited on page 165).

[Bre16]   Brehmer, M. M. "Why Visualization? Task Abstraction for Analysis and Design". PhD thesis. University of British Columbia, 2016 (cited on page 50).

[BRL09]   Bertini, E., Rigamonti, M., and Lalanne, D. "Extended Excentric Labeling". In: *Computer Graphics Forum* 28.3 (2009), pp. 927–934. DOI: 10 . 1111 / j . 1467 - 8659 . 2009 . 01456 . x (cited on pages 175, 177).

[BRT95]   Bergman, L. D., Rogowitz, B. E., and Treinish, L. "A Rule-Based Tool for Assisting Colormap Selection". In: *Proceedings of the IEEE Visualization Conference (Vis).* IEEE Computer Society, 1995, pp. 118–125. DOI: 10.1109/VISUAL.1995.480803 (cited on pages 56, 127).

[Buj+91]   Buja, A., McDonald, J. A., Michalak, J., and Stuetzle, W. "Interactive Data Visualization Using Focusing and Linking". In: *Proceedings of the IEEE Visualization Conference (Vis).* IEEE Computer Society, 1991, pp. 156–163, 419. DOI: 10 . 1109 / VISUAL . 1991 . 175794 (cited on page 157).

[But+08]   Butkiewicz, T., Dou, W., Wartell, Z., Ribarsky, W., and Chang, R. "Multi-Focused Geospatial Analysis Using Probes". In: *IEEE Transactions on Visualization and Computer Graphics* 14.6 (2008), pp. 1165–1172. DOI: 10 . 1109 / TVCG . 2008 . 149 (cited on page 102).

[Bux90]   Buxton, W. "A Three-state Model of Graphical Input". In: *Proceedings of the IFIP International Conference on Human-Computer Interaction (INTERACT).* North-Holland, 1990, pp. 449–456 (cited on page 145).

[BW08a]   Bachthaler, S. and Weiskopf, D. "Continuous Scatterplots". In: *IEEE Transactions on Visualization and Computer Graphics* 14.6 (2008), pp. 1428–1435. DOI: 10.1109/TVCG.2008.119 (cited on pages 209, 265, 345).

[BW08b]   Byron, L. and Wattenberg, M. "Stacked Graphs – Geometry & Aesthetics". In: *IEEE Transactions on Visualization and Computer Graphics* 14.6 (2008), pp. 1245–1252. DOI: 10.1109/TVCG. 2008.166 (cited on page 89).

[BW18]   Belmonte, N. and Wang, Y. *Refolding the Earth: Interactive Myriahedral Projection and Fabrication*. Poster at the IEEE Conference on Information Visualization. Berlin, Germany, 2018 (cited on page 100).

[CD04]   Cook, D. and Das, S. K. *Smart Environments: Technology, Protocols and Applications*. Wiley-Interscience, 2004. DOI: 10.1002/ 047168659X (cited on pages 269, 304).

[Cen+17] Ceneda, D., Gschwandtner, T., May, T., Miksch, S., Schulz, H.-J., Streit, M., and Tominski, C. "Characterizing Guidance in Visual Analytics". In: *IEEE Transactions on Visualization and Computer Graphics* 23.1 (2017), pp. 111–120. DOI: 10.1109/TVCG.2016. 2598468 (cited on pages 277–279, 283, 304).

[Che04]   Chen, H. "Compound Brushing Explained". In: *Information Visualization* 3.2 (2004), pp. 96–108. DOI: 10.1057/palgrave.ivs. 9500068 (cited on page 158).

[Chi00]   Chi, E. H.-H. "A Taxonomy of Visualization Techniques Using the Data State Reference Model". In: *Proceedings of the IEEE Symposium Information Visualization (InfoVis)*. IEEE Computer Society, 2000, pp. 69–75. DOI: 10.1109/INFVIS.2000.885092 (cited on page 45).

[Chu+15] Chung, H., North, C., Joshi, S., and Chen, J. "Four Considerations for Supporting Visual Analysis in Display Ecologies". In: *Proceedings of the IEEE Conference on Visual Analytics Science and Technology (VAST)*. IEEE Computer Society, 2015, pp. 33–40. DOI: 10.1109/VAST.2015.7347628 (cited on page 304).

[CKB08]   Cockburn, A., Karlson, A., and Bederson, B. B. "A Review of Overview+Detail, Zooming, and Focus+Context Interfaces". In: *ACM Computing Surveys* 41.1 (2008), 2:1–2:31. DOI: 10.1145/ 1456650.1456652 (cited on page 64).

[Cle93]   Cleveland, W. S. *Visualizing Data*. Hobart Press, 1993 (cited on pages 69, 91).

[CM84]   Cleveland, W. S. and McGill, R. "Graphical Perception: Theory, Experimentation, and Application to the Development of Graphical Methods". In: *Journal of the American Statistical Association* 79.387 (1984), pp. 531–554. DOI: 10.1080/01621459.1984. 10478080 (cited on page 55).

[Col+18] Collins, C., Andrienko, N., Schreck, T., Yang, J., Choo, J., Engelke, U., Jena, A., and Dwyer, T. "Guidance in the Human-Machine Analytics Process". In: *Visual Informatics* 3.1 (2018). DOI: 10.1016/j.visinf.2018.09.003 (cited on pages 277, 304).

[Con+08] Conversy, S., Barboni, E., Navarre, D., and Palanque, P. "Improving Modularity of Interactive Software with the MDPC Architecture". In: *Engineering Interactive Systems: EIS 2007 Joint Working Conferences, EHCI 2007, DSV-IS 2007, HCSE 2007, Salamanca, Spain, March 22-24, 2007. Selected Papers.* Edited by Gulliksen, J., Harning, M. B., Palanque, P., van der Veer, G. C., and Wesson, J. Springer, 2008, pp. 321–338. DOI: 10.1007/978-3-540-92698-6_20 (cited on page 196).

[CR98] Chi, E. H.-H. and Riedl, J. T. "An Operator Interaction Framework for Visualization Systems". In: *Proceedings of the IEEE Symposium Information Visualization (InfoVis)*. IEEE Computer Society, 1998, pp. 63–70. DOI: 10.1109/INFVIS.1998.729560 (cited on page 44).

[CRC07] Cooper, A., Reimann, R., and Cronin, D. *About Face 3: The Essentials of Interaction Design*. Wiley, 2007 (cited on pages 139, 140).

[CvW11] Claessen, J. H. and van Wijk, J. J. "Flexible Linked Axes for Multivariate Data Visualization". In: *IEEE Transactions on Visualization and Computer Graphics* 17.12 (2011), pp. 2310–2316. DOI: 10.1109/TVCG.2011.201 (cited on page 72).

[DGH03] Doleisch, H., Gasser, M., and Hauser, H. "Interactive Feature Specification for Focus+Context Visualization of Complex Simulation Data". In: *Proceedings of the Joint Eurographics - IEEE TCVG Symposium on Visualization (VisSym)*. Eurographics Association, 2003, pp. 239–248. DOI: 10.2312/VisSym/VisSym03/239-248 (cited on pages 221, 265).

[DH02a] Doleisch, H. and Hauser, H. "Smooth Brushing for Focus+Context Visualization of Simulation Data in 3D". In: *Journal of WSCG* 10.1-3 (2002), pp. 147–154. URL: http://wscg.zcu.cz/wscg2002/Papers_2002/E71.pdf (cited on page 156).

[DH02b] Dragicevic, P. and Huot, S. "SpiraClock: A Continuous and Non-Intrusive Display for Upcoming Events". In: *Proceedings of the SIGCHI Conference Human Factors in Computing Systems (CHI)*. Extended Abstracts. ACM Press, 2002, pp. 604–605. DOI: 10.1145/506443.506505 (cited on page 195).

[Dix+04] Dix, A., Finlay, J., Abowd, G. D., and Beale, R. *Human-Computer Interaction*. 3rd edition. Pearson Education, 2004 (cited on page 206).

[DP20]   Dimara, E. and Perin, C. "What is Interaction for Data Visualization?" In: *IEEE Transactions on Visualization and Computer Graphics* 26.1 (2020), pp. 119–129. DOI: 10.1109/TVCG.2019.2934283 (cited on page 206).

[dSai39]   De Saint-Exupéry, A. *Wind, Sand, and Stars.* translated by Lewis Galantière. Harcourt, Inc., 1939 (cited on page 62).

[Düb+14]   Dübel, S., Röhlig, M., Schumann, H., and Trapp, M. "2D and 3D Presentation of Spatial Data: A Systematic Review". In: *InfoVis Workshop: Does 3D Really Make Sense for Data Visualization?* IEEE Computer Society, 2014. DOI: 10.1109/3DVis.2014.7160094 (cited on pages 103, 104).

[Düb+17]   Dübel, S., Röhlig, M., Tominski, C., and Schumann, H. "Visualizing 3D Terrain, Geo-Spatial Data, and Uncertainty". In: *Informatics* 4.1 (2017), pp. 1–18. DOI: 10.3390/informatics4010006 (cited on page 126).

[ED06]   Ellis, G. and Dix, A. J. "The Plot, the Clutter, the Sampling and its Lens: Occlusion Measures for Automatic Clutter Reduction". In: *Proceedings of the Conference on Advanced Visual Interfaces (AVI).* ACM Press, 2006, pp. 266–269. DOI: 10.1145/1133265.1133318 (cited on pages 178, 265).

[EF10]   Elmqvist, N. and Fekete, J.-D. "Hierarchical Aggregation for Information Visualization: Overview, Techniques, and Design Guidelines". In: *IEEE Transactions on Visualization and Computer Graphics* 16.3 (2010), pp. 439–454. DOI: 10.1109/TVCG.2009.84 (cited on pages 233, 265).

[Eic+14]   Eichner, C., Bittig, A., Schumann, H., and Tominski, C. "Analyzing Simulations of Biochemical Systems with Feature-Based Visual Analytics." In: *Computers & Graphics* 38.1 (2014), pp. 18–26. DOI: 10.1016/j.cag.2013.09.001 (cited on page 221).

[Eic+15]   Eichner, C., Nocke, T., Schulz, H. J., and Schumann, H. "Interactive Presentation of Geo-Spatial Climate Data in Multi-Display Environments". In: *ISPRS International Journal of Geo-Information* 4.2 (2015), pp. 493–514. DOI: 10.3390/ijgi4020493 (cited on pages 272, 276).

[Eic94]   Eick, S. G. "Data Visualization Sliders". In: *Proceedings of the ACM Symposium on User Interface Software and Technology (UIST).* ACM Press, 1994, pp. 119–120. DOI: 10.1145/192426.192472 (cited on page 152).

[Elm+11]   Elmqvist, N., Moere, A. V., Jetter, H.-C., Cernea, D., Reiterer, H., and Jankun-Kelly, T. "Fluid Interaction for Information Visualization". In: *Information Visualization* 10.4 (2011), pp. 327–340. DOI: 10.1177/1473871611413180 (cited on page 143).

[Emm+16]   Emmons, S., Kobourov, S., Gallant, M., and Börner, K. "Analysis of Network Clustering Algorithms and Cluster Quality Metrics at Scale". In: *PLOS ONE* 11.7 (2016), pp. 1–18. DOI: 10.1371/journal.pone.0159161 (cited on page 265).

[End+17]   Endert, A., Ribarsky, W., Turkay, C., Wong, W., Nabney, I. T., Blanco, I. D., and Rossi, F. "The State of the Art in Integrating Machine Learning into Visual Analytics". In: *Computer Graphics Forum* 36.8 (2017), pp. 458–486. DOI: 10.1111/cgf.13092 (cited on page 265).

[ENS15]   Eichner, C., Nyolt, M., and Schumann, H. "A Novel Infrastructure for Supporting Display Ecologies". In: *Advances in Visual Computing: Proceedings of the International Symposium on Visual Computing (ISVC)*. Springer, 2015, pp. 722–732. DOI: 10.1007/978-3-319-27863-6_68 (cited on pages 269, 346).

[EST19]   Eichner, C., Schumann, H., and Tominski, C. *Multi-display Visual Analysis: Model, Interface, and Layout Computation*. Tech. rep. arXiv:1912.08558 [cs.GR]. CoRR, 2019. URL: https://arxiv.org/abs/1912.08558 (cited on page 304).

[FB90]   Feiner, S. K. and Beshers, C. "Worlds within Worlds: Metaphors for Exploring n-dimensional Virtual Worlds". In: *Proceedings of the ACM Symposium on User Interface Software and Technology (UIST)*. ACM Press, 1990, pp. 76–83. DOI: 10.1145/97924.97933 (cited on page 79).

[FB95]   Furnas, G. W. and Bederson, B. B. "Space-Scale Diagrams: Understanding Multiscale Interfaces". In: *Proceedings of the SIGCHI Conference Human Factors in Computing Systems (CHI)*. ACM Press, 1995, pp. 234–241. DOI: 10.1145/223904.223934 (cited on page 206).

[FD13]   Frisch, M. and Dachselt, R. "Visualizing Offscreen Elements of Node-Link Diagrams". In: *Information Visualization* 12.2 (2013), pp. 133–162. DOI: 10.1177/1473871612473589 (cited on page 165).

[Fek+19]   Fekete, J.-D., Fisher, D., Nandi, A., and Sedlmair, M. "Progressive Data Analysis and Visualization (Dagstuhl Seminar 18411)". In: *Dagstuhl Reports* 8.10 (2019), pp. 1–40. DOI: 10.4230/DagRep.8.10.1 (cited on page 304).

[Fis+12]   Fisher, D., Popov, I. O., Drucker, S. M., and m.c. schraefel. "Trust Me, I'm Partially Right: Incremental Visualization Lets Analysts Explore Large Datasets Faster". In: *Proceedings of the SIGCHI Conference Human Factors in Computing Systems (CHI)*. ACM Press, 2012, pp. 1673–1682. DOI: 10.1145/2207676.2208294 (cited on page 304).

[FR91]   Fruchterman, T. M. J. and Reingold, E. M. "Graph Drawing by Force-Directed Placement". In: *Software: Practice and Experience* 21.11 (1991), pp. 1129–1164. DOI: 10.1002/spe.4380211102 (cited on page 113).

[Fra98]   Frank, A. U. "Different Types of "Times" in GIS". In: *Spatial and Temporal Reasoning in Geographic Information Systems*. Edited by Egenhofer, M. J. and Golledge, R. G. Oxford University Press, 1998, pp. 40–62 (cited on page 84).

[FS04]   Fuchs, G. and Schumann, H. "Intelligent Icon Positioning for Interactive Map-Based Information Systems". In: *Innovations Through Information Technology*. Edited by Khosrow-Pour, M. Hershey, PA, USA: Idea Group Inc., 2004, pp. 261–264. DOI: 10.4018/978-1-59140-261-9.ch067. URL: http://www.irma-international.org/viewtitle/32349/ (cited on page 102).

[Fur86]   Furnas, G. W. "Generalized Fisheye Views". In: *Proceedings of the SIGCHI Conference Human Factors in Computing Systems (CHI)*. ACM Press, 1986, pp. 16–23. DOI: 10.1145/22339.22342 (cited on page 214).

[Fur97]   Furnas, G. W. "Effective View Navigation". In: *Proceedings of the SIGCHI Conference Human Factors in Computing Systems (CHI)*. ACM Press, 1997, pp. 367–374. DOI: 10.1145/258549.258800 (cited on page 206).

[Gla+14]   Gladisch, S., Schumann, H., Ernst, M., Füllen, G., and Tominski, C. "Semi-Automatic Editing of Graphs with Customized Layouts". In: *Computer Graphics Forum* 33.3 (2014), pp. 381–390. DOI: 10.1111/cgf.12394 (cited on pages 182, 183).

[Gla14]   Glaßer, S. "Visual Analysis, Clustering, and Classification of Contrast-Enhanced Tumor Perfusion MRI Data". PhD thesis. Otto von Guericke University Magdeburg, 2014 (cited on page 247).

[Gle+11]   Gleicher, M., Albers, D., Walker, R., Jusufi, I., Hansen, C. D., and Roberts, J. C. "Visual Comparison for Information Visualization". In: *Information Visualization* 10.4 (2011), pp. 289–309. DOI: 10.1177/1473871611416549 (cited on pages 184, 206).

[Gle18]   Gleicher, M. "Considerations for Visualizing Comparison". In: *IEEE Transactions on Visualization and Computer Graphics* 24.1 (2018), pp. 413–423. DOI: 10.1109/TVCG.2017.2744199 (cited on page 206).

[GS06]   Griethe, H. and Schumann, H. "The Visualization of Uncertain Data: Methods and Problems". In: *Proceedings of the Simulation and Visualization (SimVis)*. SCS Publishing House e.V., 2006, pp. 143–156 (cited on page 125).

[GST13]   Gladisch, S., Schumann, H., and Tominski, C. "Navigation Recommendations for Exploring Hierarchical Graphs". In: *Advances in Visual Computing: Proceedings of the International Symposium on Visual Computing (ISVC)*. Springer, 2013, pp. 36–47. DOI: 10.1007/978-3-642-41939-3_4 (cited on pages 165, 283).

[Gua13]   Guastello, S. J. *Human Factors Engineering and Ergonomics: A Systems Approach.* 2nd edition. CRC Press, 2013 (cited on page 50).

[Gus+08]   Gustafson, S., Baudisch, P., Gutwin, C., and Irani, P. "Wedge: Clutter-Free Visualization of Off-Screen Locations". In: *Proceedings of the SIGCHI Conference Human Factors in Computing Systems (CHI)*. ACM Press, 2008, pp. 787–796. DOI: 10.1145/1357054.1357179 (cited on page 165).

[GYZ14]   Gross, J. L., Yellen, J., and Zhang, P., eds. *Handbook of Graph Theory.* CRC Press, 2014 (cited on page 112).

[Had+10]   Hadlak, S., Tominski, C., Schulz, H.-J., and Schumann, H. "Visualization of Attributed Hierarchical Structures in a Spatio-Temporal Context". In: *International Journal of Geographical Information Science* 24.10 (2010), pp. 1497–1513. DOI: 10.1080/13658816.2010.510840 (cited on page 121).

[Had+13]   Hadlak, S., Schumann, H., Cap, C. H., and Wollenberg, T. "Supporting the Visual Analysis of Dynamic Networks by Clustering Associated Temporal Attributes". In: *IEEE Transactions on Visualization and Computer Graphics* 19.12 (2013), pp. 2267–2276. DOI: 10.1109/TVCG.2013.198 (cited on page 253).

[Had14]   Hadlak, S. "Graph Visualization in Space and Time". PhD thesis. University of Rostock, 2014 (cited on pages 251, 252, 254).

[Häg70]   Hägerstrand, T. "What About People in Regional Science?" In: *Papers of the Regional Science Association* 24 (1970), pp. 7–21 (cited on page 108).

[Hal+16]   Hall, K. W., Perin, C., Kusalik, P. G., Gutwin, C., and Carpendale, M. S. T. "Formalizing Emphasis in Information Visualization". In: *Computer Graphics Forum* 35.3 (2016), pp. 717–737. DOI: 10.1111/cgf.12936 (cited on page 155).

[Han09]   Hanrahan, P. *Systems of Thought.* Keynote presentation at the Eurographics/IEEE Symposium on Visualization (EuroVis). 2009 (cited on page 52).

[Har96]   Harris, R. L. *Information Graphics: A Comprehensive Illustrated Reference.* Managment Graphics, 1996 (cited on page 72).

[Hau06]   Hauser, H. "Generalizing Focus+Context Visualization". In: *Scientific Visualization: The Visual Extraction of Knowledge from Data*. Springer, 2006, pp. 305–327. DOI: 10.1007/3-540-30790-7_18 (cited on page 64).

[HAW08]   Heer, J., Agrawala, M., and Willett, W. "Generalized Selection via Interactive Query Relaxation". In: *Proceedings of the SIGCHI Conference Human Factors in Computing Systems (CHI)*. ACM Press, 2008, pp. 959–968. DOI: 10.1145/1357054.1357203 (cited on page 158).

[HB03]   Harrower, M. A. and Brewer, C. A. "ColorBrewer.org: An Online Tool for Selecting Color Schemes for Maps". In: *The Cartographic Journal* 40.1 (2003), pp. 27–37. DOI: 10.1179/000870403235002042 (cited on page 127).

[HB10]   Heer, J. and Bostock, M. "Crowdsourcing Graphical Perception: Using Mechanical Turk to Assess Visualization Design". In: *Proceedings of the SIGCHI Conference Human Factors in Computing Systems (CHI)*. ACM Press, 2010, pp. 203–212. DOI: 10.1145/1753326.1753357 (cited on page 55).

[HC04]   Heer, J. and Card, S. K. "DOITrees Revisited: Scalable, Space-Constrained Visualization of Hierarchical Data". In: *Proceedings of the Conference on Advanced Visual Interfaces (AVI)*. ACM Press, 2004, pp. 421–424. DOI: 10.1145/989863.989941 (cited on page 215).

[HE12]   Healey, C. G. and Enns, J. T. "Attention and Visual Memory in Visualization and Computer Graphics". In: *IEEE Transactions on Visualization and Computer Graphics* 18.7 (2012), pp. 1170–1188. DOI: 10.1109/TVCG.2011.127 (cited on pages 56, 156).

[Hei+12]   Heinrich, J., Luo, Y., Kirkpatrick, A. E., and Weiskopf, D. "Evaluation of a Bundling Technique for Parallel Coordinates". In: *Proceedings of the International Conference on Computer Graphics Theory and Applications and International Conference on Information Visualization Theory and Applications (VISIGRAPP)*. SciTePress, 2012, pp. 594–602. DOI: 10.5220/0003821205940602 (cited on page 213).

[HFM07]   Henry, N., Fekete, J.-D., and McGuffin, M. J. "NodeTrix: a Hybrid Visualization of Social Networks". In: *IEEE Transactions on Visualization and Computer Graphics* 13.6 (2007), pp. 1302–1309. DOI: 10.1109/TVCG.2007.70582 (cited on page 118).

[HHN85]   Hutchins, E. L., Hollan, J. D., and Norman, D. A. "Direct Manipulation Interfaces". In: *Human-Computer Interaction* 1.4 (1985), pp. 311–338. DOI: 10.1207/s15327051hci0104_2 (cited on pages 138, 194, 205).

[HJ05]   Hansen, C. D. and Johnson, C. R., eds. *The Visualization Handbook*. Elsevier, 2005 (cited on pages 21, 22).

[HKP11]   Han, J., Kamber, M., and Pei, J. *Data Mining: Concepts and Techniques*. Morgan Kaufmann, 2011 (cited on pages 244, 265).

[HM90]   Haber, R. B. and McNabb, D. A. "Visualization Idioms: A Conceptual Model for Scientific Visualization Systems". In: *Visualization in Scientific Computing*. Edited by Nielson, G. M., Shriver, B. D., and Rosenblum, L. J. IEEE Computer Society, 1990, pp. 74–93 (cited on page 44).

[Hol06]   Holten, D. "Hierarchical Edge Bundles: Visualization of Adjacency Relations in Hierarchical Data". In: *IEEE Transactions on Visualization and Computer Graphics* 12.5 (2006), pp. 741–748. DOI: 10.1109/TVCG.2006.147 (cited on page 213).

[Hor+19]   Horak, T., Mathisen, A., Klokmose, C. N., Dachselt, R., and Elmqvist, N. "Vistribute: Distributing Interactive Visualizations in Dynamic Multi-Device Setups". In: *Proceedings of the SIGCHI Conference Human Factors in Computing Systems (CHI)*. ACM Press, 2019, 616:1–616:13. DOI: 10.1145/3290605.3300846 (cited on page 303).

[Hor99]   Horvitz, E. "Principles of Mixed-Initiative User Interfaces". In: *Proceedings of the SIGCHI Conference Human Factors in Computing Systems (CHI)*. ACM Press, 1999, pp. 159–166. DOI: 10.1145/302979.303030 (cited on page 304).

[HS12]   Heer, J. and Shneiderman, B. "Interactive Dynamics for Visual Analysis". In: *Communications of the ACM* 55.4 (2012), pp. 45–54. DOI: 10.1145/2133806.2133821 (cited on pages 144, 289).

[HSS15]   Hadlak, S., Schumann, H., and Schulz, H.-J. "A Survey of Multi-faceted Graph Visualization". In: *EuroVis State of the Art Reports*. Eurographics Association, 2015, pp. 1–20. DOI: 10.2312/eurovisstar.20151109 (cited on pages 112, 119, 120).

[HvW09]   Holten, D. and van Wijk, J. J. "Force-Directed Edge Bundling for Graph Visualization". In: *Computer Graphics Forum* 28.3 (2009), pp. 983–990. DOI: 10.1111/j.1467-8659.2009.01450.x (cited on page 213).

[II91]   II, R. J. M. *Introduction to Shannon Sampling and Interpolation Theory*. Springer, 1991 (cited on page 232).

[Ins09]   Inselberg, A. *Parallel Coordinates – Visual Multidimensional Geometry and Its Applications*. Springer, 2009. DOI: 10.1007/978-0-387-68628-8 (cited on page 71).

[Ise+13]   Isenberg, P., Isenberg, T., Hesselmann, T., Lee, B., von Zadow, U., and Tang, A. "Data Visualization on Interactive Surfaces: A Research Agenda". In: *IEEE Computer Graphics and Applications* 33.2 (2013), pp. 16–24. DOI: 10.1109/MCG.2013.24 (cited on page 206).

[Jak+13]   Jakobsen, M. R., Haile, Y. S., Knudsen, S., and Hornbæk, K. "Information Visualization and Proxemics: Design Opportunities and Empirical Findings". In: *IEEE Transactions on Visualization and Computer Graphics* 19.12 (2013), pp. 2386–2395. DOI: 10.1109/TVCG.2013.166 (cited on page 204).

[JD13]   Jansen, Y. and Dragicevic, P. "An Interaction Model for Visualizations Beyond The Desktop". In: *IEEE Transactions on Visualization and Computer Graphics* 19.12 (2013), pp. 2396–2405. DOI: 10.1109/TVCG.2013.134 (cited on page 206).

[JMG07]   Jankun-Kelly, T. J., Ma, K.-L., and Gertz, M. "A Model and Framework for Visualization Exploration". In: *IEEE Transactions on Visualization and Computer Graphics* 13.2 (2007), pp. 357–369. DOI: 10.1109/TVCG.2007.28 (cited on page 44).

[Jol02]   Jolliffe, I. T. *Principal Component Analysis.* 2nd edition. Springer, New York, USA, 2002 (cited on page 265).

[JTS08]   John, M., Tominski, C., and Schumann, H. "Visual and Analytical Extensions for the Table Lens". In: *Proceedings of the Conference on Visualization and Data Analysis (VDA).* SPIE/IS&T, 2008, pp. 680907-1–680907-12. DOI: 10.1117/12.766440 (cited on pages 57, 60, 249).

[Kei+06]   Keim, D. A., Mansmann, F., Schneidewind, J., and Ziegler, H. "Challenges in Visual Data Analysis". In: *Proceedings of the International Conference Information Visualisation (IV).* IEEE Computer Society, 2006, pp. 9–16. DOI: 10.1109/IV.2006.31 (cited on pages 3, 207, 208, 257, 264, 265).

[Kei+10]   Keim, D., Kohlhammer, J., Ellis, G., and Mannsmann, F., eds. *Mastering the Information Age – Solving Problems with Visual Analytics.* Eurographics Association, 2010 (cited on page 265).

[Kim+17]   Kim, H., Choo, J., Lee, C., Lee, H., Reddy, C. K., and Park, H. "PIVE: Per-Iteration Visualization Environment for Real-Time Interactions with Dimension Reduction and Clustering". In: *Proceedings of the Thirty-First AAAI Conference on Artificial Intelligence.* AAAI Press, 2017, pp. 1001–1009. URL: http://aaai.org/ocs/index.php/AAAI/AAAI17/paper/view/14381 (cited on page 298).

[Kis+17]   Kister, U., Klamka, K., Tominski, C., and Dachselt, R. "GraSp: Combining Spatially-aware Mobile Devices and a Display Wall for Graph Visualization and Interaction". In: *Computer Graphics Forum* 36.3 (2017), pp. 503–514. DOI: 10.1111/cgf.13206 (cited on page 303).

[KK17]   Kerracher, N. and Kennedy, J. "Constructing and Evaluating Visualisation Task Classifications: Process and Considerations". In: *Computer Graphics Forum* 36.3 (2017), pp. 47–59. DOI: 10.1111/cgf.13167 (cited on pages 31, 50).

[KK93]   Keller, P. R. and Keller, M. M. *Visual Cues: Practical Data Visualization.* IEEE Computer Society, 1993 (cited on page 34).

[KK94]   Keim, D. A. and Kriegel, H.-P. "VisDB: Database Exploration Using Multidimensional Visualization". In: *IEEE Computer Graphics and Applications* 14.5 (1994), pp. 40–49. DOI: 10.1109/38.310723 (cited on page 75).

[KM94]   Kirsh, D. and Maglio, P. "On Distinguishing Epistemic from Pragmatic Action". In: *Cognitive Science* 18.4 (1994), pp. 513–549. DOI: 10.1207/s15516709cog1804_1 (cited on page 131).

[KNS04]   Kreuseler, M., Nocke, T., and Schumann, H. "A History Mechanism for Visual Data Mining". In: *Proceedings of the IEEE Symposium Information Visualization (InfoVis)*. IEEE Computer Society, 2004, pp. 49–56. DOI: 10.1109/INFVIS.2004.2 (cited on page 274).

[KP88]   Krasner, G. E. and Pope, S. T. "A Cookbook for Using the Model-View-Controller User Interface Paradigm in Smalltalk-80". In: *Journal of Object-Oriented Programming* 1.3 (1988), pp. 26–49 (cited on page 147).

[KPW14]   Kerren, A., Purchase, H. C., and Ward, M. O., eds. *Multivariate Network Visualization.* Springer, 2014. DOI: 10.1007/978-3-319-06793-3 (cited on page 127).

[Kra03]   Kraak, M.-J. "The Space-Time Cube Revisited from a Geovisualization Perspective". In: *Proceedings of the 21st International Cartographic Conference (ICC)*. Newcastle, UK: The International Cartographic Association (ICA), 2003, pp. 1988–1995 (cited on page 108).

[KRD14]   Kister, U., Reipschläger, P., and Dachselt, R. "Multi-Touch Manipulation of Magic Lenses for Information Visualization". In: *Proceedings of the International Conference on Interactive Tabletops and Surfaces (ITS)*. ACM Press, 2014, pp. 431–434. DOI: 10.1145/2669485.2669528 (cited on page 178).

[KS99] Kreuseler, M. and Schumann, H. "Information Visualization Using a New Focus+Context Technique in Combination with Dynamic Clustering of Information Space". In: *Proceedings of the ACM Workshop on New Paradigms in Information Visualization and Manipulation (NPIVM)*. ACM Press, 1999, pp. 1–5 (cited on page 123).

[Kul06] Kulpa, Z. "A Diagrammatic Approach to Investigate Interval Relations". In: *Journal of Visual Languages & Computing* 17.5 (2006), pp. 466–502. DOI: 10.1016/j.jvlc.2005.10.004. URL: http://dx.doi.org/10.1016/j.jvlc.2005.10.004 (cited on page 88).

[LA94] Leung, Y. K. and Apperley, M. D. "A Review and Taxonomy of Distortion-Oriented Presentation Techniques". In: *ACM Transactions on Computer-Human Interaction* 1.2 (1994), pp. 126–160. DOI: 10.1145/180171.180173 (cited on page 64).

[Lam08] Lam, H. "A Framework of Interaction Costs in Information Visualization". In: *IEEE Transactions on Visualization and Computer Graphics* 14.6 (2008), pp. 1149–1156. DOI: 10.1109/TVCG.2008.109 (cited on page 136).

[Lee+12] Lee, B., Isenberg, P., Riche, N. H., and Carpendale, S. "Beyond Mouse and Keyboard: Expanding Design Considerations for Information Visualization Interactions". In: *IEEE Transactions on Visualization and Computer Graphics* 18.12 (2012), pp. 2689–2698. DOI: 10.1109/TVCG.2012.204 (cited on page 206).

[Leh+11] Lehmann, A., Schumann, H., Staadt, O., and Tominski, C. "Physical Navigation to Support Graph Exploration on a Large High-Resolution Display". In: *Advances in Visual Computing: Proceedings of the International Symposium on Visual Computing (ISVC)*. Springer, 2011, pp. 496–507. DOI: 10.1007/978-3-642-24028-7_46 (cited on pages 202–204).

[Lex+12] Lex, A., Streit, M., Schulz, H.-J., Partl, C., Schmalstieg, D., Park, P. J., and Gehlenborg, N. "StratomeX: Visual Analysis of Large-Scale Heterogeneous Genomics Data for Cancer Subtype Characterization". In: *Computer Graphics Forum* 31.3 (2012), pp. 1175–1184. DOI: 10.1111/j.1467-8659.2012.03110.x (cited on page 247).

[LF06] Leskovec, J. and Faloutsos, C. "Sampling from Large Graphs". In: *Proceedings of the ACM Conference on Knowledge discovery and data mining (SIGKDD)*. ACM Press, 2006, pp. 631–636. DOI: 10.1145/1150402.1150479 (cited on page 232).

[LH14]   Liu, Z. and Heer, J. "The Effects of Interactive Latency on Exploratory Visual Analysis". In: *IEEE Transactions on Visualization and Computer Graphics* 20.12 (2014), pp. 2122–2131. DOI: 10.1109/TVCG.2014.2346452 (cited on pages 140, 289).

[LHT17]   Lhuillier, A., Hurter, C., and Telea, A. "State of the Art in Edge and Trail Bundling Techniques". In: *Computer Graphics Forum* 36.3 (2017), pp. 619–645. DOI: 10.1111/cgf.13213 (cited on pages 212, 213, 265).

[Lia05]   Liao, W. "Clustering of Time Series Data—A Survey". In: *Pattern Recognition* 38.11 (2005), pp. 1857–1874. DOI: 10.1016/j.patcog.2005.01.025 (cited on page 251).

[LKD19]   Langner, R., Kister, U., and Dachselt, R. "Multiple Coordinated Views at Large Displays for Multiple Users: Empirical Findings on User Behavior, Movements, and Distances". In: *IEEE Transactions on Visualization and Computer Graphics* 25.1 (2019), pp. 608–618. DOI: 10.1109/TVCG.2018.2865235 (cited on page 303).

[Loh19]   Lohr, S. L. *Sampling: Design and Analysis*. 2nd edition. CRC Press, 2019 (cited on page 232).

[LS07]   Luboschik, M. and Schumann, H. "Explode to Explain – Illustrative Information Visualization". In: *Proceedings of the International Conference Information Visualisation (IV)*. IEEE Computer Society, 2007. DOI: 10.1109/IV.2007.50 (cited on pages 77, 345).

[LTM18]   Lam, H., Tory, M., and Munzner, T. "Bridging from Goals to Tasks with Design Study Analysis Reports". In: *IEEE Transactions on Visualization and Computer Graphics* 24.1 (2018), pp. 435–445. DOI: 10.1109/TVCG.2017.2744319 (cited on pages 29, 50).

[Lub+12]   Luboschik, M., Maus, C., Schulz, H.-J., Schumann, H., and Uhrmacher, A. "Heterogeneity-Based Guidance for Exploring Multiscale Data in Systems Biology". In: *Proceedings of the IEEE Symposium on Biological Data Visualization (BioVis)*. IEEE Computer Society, 2012, pp. 33–40. DOI: 10.1109/BioVis.2012.6378590 (cited on pages 233, 237, 265).

[Lub+15]   Luboschik, M., Röhlig, M., Bittig, A. T., Andrienko, N., Schumann, H., and Tominski, C. "Feature-Driven Visual Analytics of Chaotic Parameter-Dependent Movement". In: *Computer Graphics Forum* 34.3 (2015), pp. 421–430. DOI: 10.1111/cgf.12654 (cited on pages 225, 229).

[LV07]   Lee, J. A. and Verleysen, M. *Nonlinear Dimensionality Reduction*. Springer, 2007. DOI: 10.1007/978-0-387-39351-3 (cited on page 265).

[LWW90]   LeBlanc, J., Ward, M. O., and Wittels, N. "Exploring n-Dimensional Databases". In: *Proceedings of the IEEE Visualization Conference (Vis)*. IEEE Computer Society, 1990, pp. 230–237. DOI: 10.1109/VISUAL.1990.146386 (cited on page 78).

[Mac86]   Mackinlay, J. "Automating the Design of Graphical Presentations of Relational Information". In: *ACM Transactions on Graphics* 5.2 (1986), pp. 110–141. DOI: 10.1145/22949.22950 (cited on pages 50, 54).

[Mac95]   MacEachren, A. M. *How Maps Work: Representation, Visualization, and Design*. Guilford Press, 1995 (cited on pages 54, 99, 127).

[Mar+18]   Marriott, K., Schreiber, F., Dwyer, T., Klein, K., Riche, N. H., Itoh, T., Stuerzlinger, W., and Thomas, B. H., eds. *Immersive Analytics*. Springer, 2018. DOI: 10.1007/978-3-030-01388-2 (cited on page 206).

[MDB87]   McCormick, B. H., DeFanti, T. A., and Brown, M. D. "Visualization in Scientific Computing". In: *ACM SIGGRAPH Computer Graphics* 21.6 (1987), p. 3. DOI: 10.1145/41997.41998 (cited on page 2).

[Mey+15]   Meyer, M., Sedlmair, M., Quinan, P. S., and Munzner, T. "The Nested Blocks and Guidelines Model". In: *Information Visualization* 14.3 (2015), pp. 234–249. DOI: 10.1177/1473871613510429 (cited on page 41).

[ML17]   McNabb, L. and Laramee, R. S. "Survey of Surveys (SoS) - Mapping The Landscape of Survey Papers in Information Visualization". In: *Computer Graphics Forum* 36.3 (2017), pp. 589–617. DOI: 10.1111/cgf.13212 (cited on page 126).

[MNS06]   Müller, W., Nocke, T., and Schumann, H. "Enhancing the Visualization Process with Principal Component Analysis to Support the Exploration of Trends". In: *Asia-Pacific Symposium on Information Visualisation (APVIS)*. Australian Computer Society, 2006, pp. 121–130. URL: https://dl.acm.org/citation.cfm?id=1151922 (cited on page 261).

[Mos+09]   Moscovich, T., Chevalier, F., Henry, N., Pietriga, E., and Fekete, J.-D. "Topology-Aware Navigation in Large Networks". In: *Proceedings of the SIGCHI Conference Human Factors in Computing Systems (CHI)*. ACM Press, 2009, pp. 2319–2328. DOI: 10.1145/1518701.1519056 (cited on page 166).

[Müh+14] Mühlbacher, T., Piringer, H., Gratzl, S., Sedlmair, M., and Streit, M. "Opening the Black Box: Strategies for Increased User Involvement in Existing Algorithm Implementations". In: *IEEE Transactions on Visualization and Computer Graphics* 20.12 (2014), pp. 1643–1652. DOI: 10.1109/TVCG.2014.2346578 (cited on page 304).

[Mun09] Munzner, T. "A Nested Model for Visualization Design and Validation". In: *IEEE Transactions on Visualization and Computer Graphics* 15.6 (2009), pp. 921–928. DOI: 10.1109/TVCG.2009.111 (cited on pages 41, 42).

[Mun14] Munzner, T. *Visualization Analysis and Design*. A K Peters/CRC Press, 2014 (cited on page 50).

[MW14] Mackinlay, J. D. and Winslow, K. *Designing Great Visualizations*. White Paper. Tableau Software Inc., 2014. URL: https://www.tableau.com/sites/default/files/media/designing-great-visualizations.pdf (cited on page 55).

[MW95] Martin, A. R. and Ward, M. O. "High Dimensional Brushing for Interactive Exploration of Multivariate Data". In: *Proceedings of the IEEE Visualization Conference (Vis)*. IEEE Computer Society, 1995, pp. 271–278. DOI: 10.1109/VISUAL.1995.485139 (cited on pages 149, 156).

[Nac+16] Nachmanson, L., Nocaj, A., Bereg, S., Zhang, L., and Holroyd, A. "Node Overlap Removal by Growing a Tree". In: *Proceedings of the International Symposium on Graph Drawing (GD)*. Springer, 2016, pp. 33–43. DOI: 10.1007/978-3-319-50106-2_3 (cited on page 114).

[NH06] Novotny, M. and Hauser, H. "Outlier-Preserving Focus+Context Visualization in Parallel Coordinates". In: *IEEE Transactions on Visualization and Computer Graphics* 12.5 (2006), pp. 893–900. DOI: 10.1109/TVCG.2006.170 (cited on pages 210, 211, 265, 346).

[Nob+19] Nobre, C., Streit, M., Meyer, M., and Lex, A. "The State of the Art in Visualizing Multivariate Networks". In: *Computer Graphics Forum* 38.3 (2019), pp. 807–832. DOI: 10.1111/cgf.13728 (cited on page 127).

[Nor13] Norman, D. A. *The Design of Everyday Things*. Revised and expanded edition. Basic Books, 2013 (cited on pages 135, 137).

[Nor88] Norman, D. A. *The Psychology of Everyday Things*. Basic Books, 1988 (cited on pages 135, 137).

[NSS05]   Nocke, T., Schlechtweg, S., and Schumann, H. "Icon-Based Visualization Using Mosaic Metaphors". In: *Proceedings of the International Conference Information Visualisation (IV)*. IEEE Computer Society, 2005, pp. 103–109. DOI: 10.1109/IV.2005.58 (cited on pages 75, 345).

[PB07]   Preim, B. and Bartz, D. *Visualization in Medicine: Theory, Algorithms, and Applications*. Morgan Kaufmann, 2007 (cited on page 22).

[Pin+12]   Pindat, C., Pietriga, E., Chapuis, O., and Puech, C. "JellyLens: Content-aware Adaptive Lenses". In: *Proceedings of the ACM Symposium on User Interface Software and Technology (UIST)*. ACM Press, 2012, pp. 261–270. DOI: 10.1145/2380116.2380150 (cited on page 177).

[Pir+09]   Piringer, H., Tominski, C., Muigg, P., and Berger, W. "A Multi-Threading Architecture to Support Interactive Visual Exploration". In: *IEEE Transactions on Visualization and Computer Graphics* 15.6 (2009), pp. 1113–1120. DOI: 10.1109/TVCG.2009.110 (cited on page 294).

[Pos+03]   Post, F. H., Vrolijk, B., Hauser, H., Laramee, R. S., and Doleisch, H. "The State of the Art in Flow Visualisation: Feature Extraction and Tracking". In: *Computer Graphics Forum* 22.4 (2003), pp. 775–792. DOI: 10.1111/j.1467-8659.2003.00723.x (cited on pages 220, 221).

[PW06]   Plumlee, M. and Ware, C. "Zooming versus Multiple Window Interfaces: Cognitive Costs of Visual Comparisons". In: *ACM Transactions on Computer-Human Interaction* 13.2 (2006), pp. 179–209. DOI: 10.1145/1165734.1165736 (cited on page 185).

[Qia+12]   Qiang, Y., Delafontaine, M., Versichele, M., Maeyer, P. D., and de Weghe, N. V. "Interactive Analysis of Time Intervals in a Two-dimensional Space". In: *Information Visualization* 11.4 (2012), pp. 255–272. DOI: 10.1177/1473871612436775 (cited on page 89).

[Rad+12]   Radloff, A., Lehmann, A., Staadt, O. G., and Schumann, H. "Smart Interaction Management: An Interaction Approach for Smart Meeting Rooms". In: *Proceedings of the Eighth International Conference on Intelligent Environments (IE)*. IEEE Computer Society, 2012, pp. 228–235. DOI: 10.1109/IE.2012.34 (cited on pages 273, 346).

[Rad+15]   Radloff, A., Tominski, C., Nocke, T., and Schumann, H. "Supporting Presentation and Discussion of Visualization Results in Smart Meeting Rooms". In: *The Visual Computer* 31.9 (2015), pp. 1271–1286. DOI: 10.1007/s00371-014-1010-x (cited on pages 274, 302, 304).

[Rag+16]   Ragan, E. D., Endert, A., Sanyal, J., and Chen, J. "Characterizing Provenance in Visualization and Data Analysis: An Organizational Framework of Provenance Types and Purposes". In: *IEEE Transactions on Visualization and Computer Graphics* 22.1 (2016), pp. 31–40. DOI: 10.1109/TVCG.2015.2467551 (cited on page 125).

[RC94]   Rao, R. and Card, S. K. "The Table Lens: Merging Graphical and Symbolic Representations in an Interactive Focus + Context Visualization for Tabular Information". In: *Proceedings of the SIGCHI Conference Human Factors in Computing Systems (CHI)*. ACM Press, 1994, pp. 318–322. DOI: 10.1145/191666.191776 (cited on page 68).

[RLP10]   Riche, N. H., Lee, B., and Plaisant, C. "Understanding Interactive Legends: a Comparative Evaluation with Standard Widgets". In: *Computer Graphics Forum* 29.3 (2010), pp. 1193–1202. DOI: 10.1111/j.1467-8659.2009.01678.x (cited on page 151).

[RLS11]   Radloff, A., Luboschik, M., and Schumann, H. "Smart Views in Smart Environments". In: *Proceedings of the Smart Graphics*. Springer, 2011, pp. 1–12. DOI: 10.1007/978-3-642-22571-0_1 (cited on pages 12, 273, 345).

[Rob07]   Roberts, J. C. "State of the Art: Coordinated & Multiple Views in Exploratory Visualization". In: *Proceedings of the International Conference on Coordinated and Multiple Views in Exploratory Visualization (CMV)*. IEEE Computer Society, 2007, pp. 98–102. DOI: 10.1109/CMV.2007.20 (cited on page 65).

[Röh+15]   Röhlig, M., Luboschik, M., Krüger, F., Kirste, T., Schumann, H., Bögl, M., Bilal, A., and Miksch, S. "Supporting Activity Recognition by Visual Analytics". In: *Proceedings of the IEEE Conference on Visual Analytics Science and Technology (VAST)*. IEEE Computer Society, 2015, pp. 41–48. DOI: 10.1109/VAST.2015.7347629 (cited on page 241).

[Ros+04]   Rosario, G. E., Rundensteiner, E. A., Brown, D. C., Ward, M. O., and Huang, S. "Mapping Nominal Values to Numbers for Effective Visualization". In: *Information Visualization* 3.2 (2004), pp. 80–95. DOI: 10.1057/palgrave.ivs.9500072 (cited on page 57).

[Ros+11] Rosenbaum, R., Giménez, A., Schumann, H., and Hamann, B. "A Flexible Low-complexity Device Adaptation Approach for Data Presentation". In: *Proceedings of the Conference on Visualization and Data Analysis (VDA)*. SPIE/IS&T, 2011, 78680F-1–78680F-12. DOI: 10.1117/12.871975 (cited on page 302).

[RPS01] Reinders, F., Post, F. H., and Spoelder, H. J. "Visualization of Time-Dependent Data with Feature Tracking and Event Detection." In: *The Visual Computer* 17.1 (2001), pp. 55–71. DOI: 10.1007/PL00013399 (cited on pages 220, 265).

[RS09] Rosenbaum, R. and Schumann, H. "Progressive Refinement: More Than a Means to Overcome Limited Bandwidth". In: *Proceedings of the Conference on Visualization and Data Analysis (VDA)*. SPIE/IS&T, 2009, pp. 72430-1–72430-13. DOI: 10.1117/12.810501 (cited on page 303).

[RS17] Röhlig, M. and Schumann, H. "Visibility Widgets for Unveiling Occluded Data in 3D Terrain Visualization". In: *Journal of Visual Languages & Computing* 42 (2017), pp. 86–98. DOI: 10.1016/j.jvlc.2017.08.008 (cited on page 105).

[Ruz+12] Ruzinoor, C. M., Shariff, A. R. M., Pradhan, B., Rodzi Ahmad, M., and Rahim, M. S. M. "A Review on 3D Terrain Visualization of GIS Data: Techniques and Software". In: *Geo-spatial Information Science* 15.2 (2012), pp. 105–115. DOI: 10.1080/10095020.2012.714101 (cited on page 101).

[Sac+14] Sacha, D., Stoffel, A., Kwon, B. C., Ellis, G., and Keim, D. A. "Knowledge Generation Model for Visual Analytics". In: *IEEE Transactions on Visualization and Computer Graphics* 20.12 (2014), pp. 1604–1613. DOI: 10.1109/TVCG.2014.2346481 (cited on pages 47, 264).

[Sac+17] Sacha, D., Zhang, L., Sedlmair, M., Lee, J., Peltonen, J., Weiskopf, D., North, S. C., and Keim, D. A. "Visual Interaction with Dimensionality Reduction: A Structured Literature Analysis". In: *IEEE Transactions on Visualization and Computer Graphics* 23.1 (2017), pp. 241–250. DOI: 10.1109/TVCG.2016.2598495 (cited on page 265).

[Sai+05] Saito, T., Miyamura, H. N., Yamamoto, M., Saito, H., Hoshiya, Y., and Kaseda, T. "Two-Tone Pseudo Coloring: Compact Visualization for One-Dimensional Data". In: *Proceedings of the IEEE Symposium Information Visualization (InfoVis)*. IEEE Computer Society, 2005, pp. 173–180. DOI: 10.1109/INFVIS.2005.1532144 (cited on page 59).

[Sar+02]   Sarwar, B., Karypis, G., Konstan, J., and Riedl, J. "Incremental Singular Value Decomposition Algorithms for Highly Scalable Recommender Systems". In: *Proceedings of the 5th International Conference on Computer and Information Technology (ICCIT).* East West University, Dhaka, Bangladesh, 2002, pp. 399–404 (cited on page 298).

[SB94]   Sarkar, M. and Brown, M. H. "Graphical Fisheye Views". In: *Communications of the ACM* 37.12 (1994), pp. 73–83. DOI: 10. 1145/198366.198384 (cited on page 178).

[Sch+13a]   Schulz, H.-J., Nocke, T., Heitzler, M., and Schumann, H. "A Design Space of Visualization Tasks". In: *IEEE Transactions on Visualization and Computer Graphics* 19.12 (2013), pp. 2366–2375. DOI: 10.1109/TVCG.2013.120 (cited on pages 29, 34, 50).

[Sch+13b]   Schulz, H.-J., Streit, M., May, T., and Tominski, C. *Towards a Characterization of Guidance in Visualization.* Poster at IEEE Conference on Information Visualization (InfoVis). Atlanta, USA, 2013 (cited on page 277).

[Sch+16]   Schulz, H.-J., Angelini, M., Santucci, G., and Schumann, H. "An Enhanced Visualization Process Model for Incremental Visualization". In: *IEEE Transactions on Visualization and Computer Graphics* 22.7 (2016), pp. 1830–1842. DOI: 10.1109/TVCG.2015. 2462356 (cited on pages 290–293, 298).

[Sch+17]   Schulz, H.-J., Nocke, T., Heitzler, M., and Schumann, H. "A Systematic View on Data Descriptors for the Visual Analysis of Tabular Data". In: *Information Visualization* 16.3 (2017), pp. 232–256. DOI: 10.1177/1473871616667767 (cited on page 26).

[Sch07]   Schaeffer, S. E. "Graph Clustering". In: *Computer Science Review* 1.1 (2007), pp. 27–64. DOI: 10.1016/j.cosrev.2007.05.001 (cited on page 250).

[SH10]   Shaer, O. and Hornecker, E. "Tangible User Interfaces: Past, Present and Future Directions". In: *Foundations and Trends in Human-Computer Interaction* 3.1–2 (2010), pp. 4–137. DOI: 10.1561/1100000026 (cited on page 197).

[She68]   Shepard, D. "A Two-dimensional Interpolation Function for Irregularly-spaced Data". In: *Proceedings of the 23rd ACM National Conference.* ACM Press, 1968, pp. 517–524. DOI: 10. 1145/800186.810616 (cited on page 25).

[Shn94]   Shneiderman, B. "Dynamic Queries for Visual Information Seeking". In: *IEEE Software* 11.6 (1994), pp. 70–77. DOI: 10.1109/ 52.329404 (cited on pages 140, 149).

[Shn96]   Shneiderman, B. "The Eyes Have It: A Task by Data Type Taxonomy for Information Visualizations". In: *Proceedings of the IEEE Symposium on Visual Languages (VL)*. IEEE Computer Society, 1996, pp. 336–343. DOI: 10.1109/VL.1996.545307 (cited on pages 34, 159).

[SHS11]   Schulz, H.-J., Hadlak, S., and Schumann, H. "The Design Space of Implicit Hierarchy Visualization: A Survey". In: *IEEE Transactions on Visualization and Computer Graphics* 17.4 (2011), pp. 393–411. DOI: 10.1109/TVCG.2010.79 (cited on page 117).

[SM00]   Schumann, H. and Müller, W. *Visualisierung: Grundlagen und Allgemeine Methoden*. Springer, 2000. DOI: 10.1007/978-3-642-57193-0 (cited on pages 13, 75, 80, 345).

[SM07]   Shen, Z. and Ma, K.-L. "Path Visualization for Adjacency Matrices". In: *Proceedings of the Joint Eurographics - IEEE TCVG Symposium on Visualization (VisSym)*. Eurographics Association, 2007, pp. 83–90. DOI: 10.2312/VisSym/EuroVis07/083-090 (cited on page 115).

[SMD12]   Spindler, M., Martsch, M., and Dachselt, R. "Going Beyond the Surface: Studying Multi-layer Interaction Above the Tabletop". In: *Proceedings of the SIGCHI Conference Human Factors in Computing Systems (CHI)*. ACM Press, 2012, pp. 1277–1286. DOI: 10.1145/2207676.2208583 (cited on page 201).

[SP09]   Shneiderman, B. and Plaisant, C. *Designing the User Interface: Strategies for Effective Human-Computer Interaction*. 5th edition. Addison-Wesley, 2009 (cited on pages 138, 143, 206).

[SP13]   Sedig, K. and Parsons, P. "Interaction Design for Complex Cognitive Activities with Visual Representations: A Pattern-Based Approach". In: *AIS Transactions on Human-Computer Interaction* 5.2 (2013), pp. 84–133 (cited on pages 133, 134).

[SP16]   Sedig, K. and Parsons, P. *Design of Visualizations for Human-Information Interaction: A Pattern-Based Framework*. Vol. 4. Synthesis Lectures on Visualization. Morgan and Claypool Publishers, 2016. DOI: 10.2200/S00685ED1V01Y201512VIS005 (cited on page 206).

[Spe07]   Spence, R. *Information Visualization: Design for Interaction*. 2nd edition. Prentice Hall, 2007 (cited on pages 13, 127, 131, 140, 162).

[SPG14]   Stolper, C. D., Perer, A., and Gotz, D. "Progressive Visual Analytics: User-Driven Visual Exploration of In-Progress Analytics". In: *IEEE Transactions on Visualization and Computer Graphics* 20.12 (2014), pp. 1653–1662. DOI: 10.1109/TVCG.2014.2346574 (cited on pages 290, 304).

[Spi+10]   Spindler, M., Tominski, C., Schumann, H., and Dachselt, R. "Tangible Views for Information Visualization". In: *Proceedings of the International Conference on Interactive Tabletops and Surfaces (ITS)*. ACM Press, 2010, pp. 157–166. DOI: 10.1145/1936652.1936684 (cited on pages 198, 199).

[Spi+14]   Spindler, M., Schuessler, M., Martsch, M., and Dachselt, R. "Pinch-Drag-Flick vs. Spatial Input: Rethinking Zoom & Pan on Mobile Displays". In: *Proceedings of the SIGCHI Conference Human Factors in Computing Systems (CHI)*. ACM Press, 2014, pp. 1113–1122. DOI: 10.1145/2556288.2557028 (cited on page 201).

[SS06]   Schulz, H.-J. and Schumann, H. "Visualizing Graphs – A Generalized View". In: *Proceedings of the International Conference Information Visualisation (IV)*. IEEE Computer Society, 2006, pp. 166–173. DOI: 10.1109/IV.2006.130 (cited on page 113).

[Ste98]   Steiner, A. "A Generalisation Approach to Temporal Data Models and their Implementations". PhD thesis. Swiss Federal Institute of Technology, Zürich, Switzerland, 1998 (cited on pages 85, 86).

[Str+12]   Streit, M., Schulz, H.-J., Lex, A., Schmalstieg, D., and Schumann, H. "Model-Driven Design for the Visual Analysis of Heterogeneous Data". In: *IEEE Transactions on Visualization and Computer Graphics* 18.6 (2012), pp. 998–1010. DOI: 10.1109/TVCG.2011.108 (cited on pages 286–288, 346).

[Tam13]   Tamassia, R., ed. *Handbook of Graph Drawing and Visualization*. CRC Press, 2013 (cited on page 113).

[TAS04]   Tominski, C., Abello, J., and Schumann, H. "Axes-Based Visualizations with Radial Layouts". In: *Proceedings of the ACM Symposium on Applied Computing (SAC)*. ACM Press, 2004, pp. 1242–1247. DOI: 10.1145/967900.968153 (cited on pages 87, 171).

[TAS09]   Tominski, C., Abello, J., and Schumann, H. "CGV – An Interactive Graph Visualization System". In: *Computers & Graphics* 33.6 (2009), pp. 660–678. DOI: 10.1016/j.cag.2009.06.002 (cited on pages 123, 166).

[Tat+12]   Tatu, A., Maass, F., Färber, I., Bertini, E., Schreck, T., Seidl, T., and Keim, D. A. "Subspace Search and Visualization to Make Sense of Alternative Clusterings in High-dimensional Data". In: *Proceedings of the IEEE Conference on Visual Analytics Science and Technology (VAST)*. IEEE Computer Society, 2012, pp. 63–72. DOI: 10.1109/VAST.2012.6400488 (cited on page 191).

[TC05]   Thomas, J. J. and Cook, K. A. *Illuminating the Path: The Research and Development Agenda for Visual Analytics*. IEEE Computer Society, 2005 (cited on page 29).

[Tei+17]   Teipel, S., Heine, C., Hein, A., Krüger, F., Kutschke, A., Kernebeck, S., Halek, M., Bader, S., and Kirste, T. "Multidimensional Assessment of Challenging Behaviors in Advanced Stages of Dementia in Nursing Homes—The insideDEM Framework". In: *Alzheimer's & Dementia: Diagnosis, Assessment & Disease Monitoring* 8 (2017), pp. 36–44. DOI: 10.1016/j.dadm.2017.03.006 (cited on page 240).

[Tel14]   Telea, A. C. *Data Visualization: Principles and Practice*. 2nd edition. A K Peters/CRC Press, 2014 (cited on pages 21, 22).

[TFJ12]   Tominski, C., Forsell, C., and Johansson, J. "Interaction Support for Visual Comparison Inspired by Natural Behavior". In: *IEEE Transactions on Visualization and Computer Graphics* 18.12 (2012), pp. 2719–2728. DOI: 10.1109/TVCG.2012.237 (cited on pages 186, 188–190).

[TFS08a]   Thiede, C., Fuchs, G., and Schumann, H. "Smart Lenses". In: *Proceedings of the Smart Graphics (SG)*. Springer, 2008, pp. 178–189. DOI: 10.1007/978-3-540-85412-8_16 (cited on page 206).

[TFS08b]   Tominski, C., Fuchs, G., and Schumann, H. "Task-Driven Color Coding". In: *Proceedings of the International Conference Information Visualisation (IV)*. IEEE Computer Society, 2008, pp. 373–380. DOI: 10.1109/IV.2008.24 (cited on page 58).

[TLH10]   Talbot, J., Lin, S., and Hanrahan, P. "An Extension of Wilkinson's Algorithm for Positioning Tick Labels on Axes". In: *IEEE Transactions on Visualization and Computer Graphics* 16.6 (2010), pp. 1036–1043. DOI: 10.1109/TVCG.2010.130 (cited on page 57).

[Tob70]   Tobler, W. R. "A Computer Movie Simulating Urban Growth in the Detroit Region". In: *Economic Geography* 46.6 (1970), pp. 234–240. DOI: 10.2307/143141 (cited on pages 25, 98).

[Tom+06]   Tominski, C., Abello, J., van Ham, F., and Schumann, H. "Fisheye Tree Views and Lenses for Graph Visualization". In: *Proceedings of the International Conference Information Visualisation (IV)*. IEEE Computer Society, 2006, pp. 17–24. DOI: 10.1109/IV.2006.54 (cited on page 179).

[Tom+12]   Tominski, C., Schumann, H., Andrienko, G., and Andrienko, N. "Stacking-Based Visualization of Trajectory Attribute Data". In: *IEEE Transactions on Visualization and Computer Graphics* 18.12 (2012), pp. 2565–2574. DOI: 10.1109/TVCG.2012.265 (cited on pages 107, 181).

[Tom+17]   Tominski, C., Gladisch, S., Kister, U., Dachselt, R., and Schumann, H. "Interactive Lenses for Visualization: An Extended Survey". In: *Computer Graphics Forum* 36.6 (2017), pp. 173–200. DOI: 10.1111/cgf.12871 (cited on pages 173, 174, 206).

[Tom15]   Tominski, C. *Interaction for Visualization*. Synthesis Lectures on Visualization 3. Morgan & Claypool, 2015. DOI: 10.2200/S00651ED1V01Y201506VIS003 (cited on pages 203, 206, 345).

[Tom16]   Tominski, C. "CompaRing: Reducing Costs of Visual Comparison". In: *Short Paper Proceedings of the Eurographics Conference on Visualization (EuroVis)*. Eurographics Association, 2016, pp. 137–141. DOI: 10.2312/eurovisshort.20161175 (cited on page 192).

[TS08]   Tominski, C. and Schumann, H. "Enhanced Interactive Spiral Display". In: *Proceedings of the Annual SIGRAD Conference, Special Theme: Interactivity*. Linköping University Electronic Press, 2008, pp. 53–56. URL: https://www.ep.liu.se/ecp_article/index.en.aspx?issue=034;article=013 (cited on page 90).

[TS12]   Tominski, C. and Schulz, H.-J. "The Great Wall of Space-Time". In: *Proceedings of the Workshop on Vision, Modeling & Visualization (VMV)*. Eurographics Association, 2012, pp. 199–206. DOI: 10.2312/PE/VMV/VMV12/199-206 (cited on page 109).

[TSS05]   Tominski, C., Schulze-Wollgast, P., and Schumann, H. "3D Information Visualization for Time Dependent Data on Maps". In: *Proceedings of the International Conference Information Visualisation (IV)*. IEEE Computer Society, 2005, pp. 175–181. DOI: 10.1109/IV.2005.3 (cited on page 108).

[Tuf83]   Tufte, E. R. *The Visual Display of Quantitative Information*. Graphics Press, 1983 (cited on pages 18, 86).

[Tuk77]   Tukey, J. W. *Exploratory Data Analysis*. Addison-Wesley, 1977 (cited on page 58).

[vHP09]   Van Ham, F. and Perer, A. "Search, Show Context, Expand on Demand: Supporting Large Graph Exploration with Degree-of-Interest ". In: *IEEE Transactions on Visualization and Computer Graphics* 15.6 (2009), pp. 953–960. DOI: 10.1109/TVCG.2009.108 (cited on page 215).

[vHSD09]   Van Ham, F., Schulz, H.-J., and Dimicco, J. M. "Honeycomb: Visual Analysis of Large Scale Social Networks". In: *Proceedings of the 12th IFIP Conference on Human-Computer Interaction (INTERACT)*. Springer, 2009, pp. 429–442. DOI: 10.1007/978-3-642-03658-3_47 (cited on page 115).

[Vic99]   Vicente, K. J. *Cognitive Work Analysis: Toward Safe, Productive, and Healthy Computer-Based Work*. CRC Press, 1999 (cited on page 50).

[vLan+11] Von Landesberger, T., Kuijper, A., Schreck, T., Kohlhammer, J., van Wijk, J. J., Fekete, J.-D., and Fellner, D. W. "Visual Analysis of Large Graphs: State-of-the-Art and Future Research Challenges". In: *Computer Graphics Forum* 30.6 (2011), pp. 1719–1749. DOI: 10.1111/j.1467-8659.2011.01898.x (cited on page 127).

[vLan+14] Von Landesberger, T., Bremm, S., Schreck, T., and Fellner, D. W. "Feature-Based Automatic Identification of Interesting Data Segments in Group Movement Data". In: *Information Visualization* 13.3 (2014), pp. 190–212. DOI: 10.1177/1473871613477851.2 (cited on page 225).

[vLan18] Von Landesberger, T. "Insights by Visual Comparison: The State and Challenges". In: *IEEE Computer Graphics and Applications* 38.3 (2018), pp. 140–148. DOI: 10.1109/MCG.2018.032421661 (cited on page 206).

[VW93] Visvalingam, M. and Whyatt, J. D. "Line Generalisation by Repeated Elimination of Points". In: *The Cartographic Journal* 30.1 (1993), pp. 46–51. DOI: 10.1179/000870493786962263 (cited on page 100).

[vWal+96] Van Walsum, T., Post, F. H., Silver, D., and Post, F. J. "Feature Extraction and Iconic Visualization". In: *IEEE Transactions on Visualization and Computer Graphics* 2.2 (1996), pp. 111–119. DOI: 10.1109/2945.506223 (cited on page 222).

[vWij06] Van Wijk, J. J. "Views on Visualization". In: *IEEE Transactions on Visualization and Computer Graphics* 12.4 (2006), pp. 421–433. DOI: 10.1109/TVCG.2006.80 (cited on pages 47, 50, 282, 289).

[vWij08] Van Wijk, J. J. "Unfolding the Earth: Myriahedral Projections". In: *The Cartographic Journal* 45.1 (2008), pp. 32–42. DOI: 10.1179/000870408X276594 (cited on page 100).

[vWN04] Van Wijk, J. J. and Nuij, W. A. A. "A Model for Smooth Viewing and Navigation of Large 2D Information Spaces". In: *IEEE Transactions on Visualization and Computer Graphics* 10.4 (2004), pp. 447–458. DOI: 10.1109/TVCG.2004.1 (cited on page 167).

[vWvS99] Van Wijk, J. J. and van Selow, E. R. "Cluster and Calendar Based Visualization of Time Series Data". In: *Proceedings of the IEEE Symposium Information Visualization (InfoVis)*. IEEE Computer Society, 1999, pp. 4–9. DOI: 10.1109/INFVIS.1999.801851 (cited on pages 92, 93, 345).

[War02] Ward, M. O. "A Taxonomy of Glyph Placement Strategies for Multidimensional Data Visualization". In: *Information Visualization* 1.2 (2002), pp. 194–210. DOI: 10.1057/palgrave.ivs.9500025 (cited on page 75).

[War12]   Ware, C. *Information Visualization: Perception for Design.* 3rd edition. Morgan Kaufmann, 2012 (cited on pages 50, 127).

[WGK15]   Ward, M. O., Grinstein, G., and Keim, D. *Interactive Data Visualization: Foundations, Techniques, and Applications.* 2nd edition. A K Peters/CRC Press, 2015 (cited on pages 13, 127).

[WH04]   Wolfe, J. M. and Horowitz, T. S. "What Attributes Guide the Deployment of Visual Attention and How do They do it?" In: *Nature Reviews Neuroscience* 05.6 (2004), pp. 495–501. DOI: 10. 1038/nrn1411 (cited on page 156).

[Wil11]   Wills, G. *Visualizing Time: Designing Graphical Representations for Statistical Data.* Springer, 2011. DOI: 10.1007/978-0-387-77907-2 (cited on page 127).

[Wil96]   Wills, G. J. "Selection: 524,288 Ways to Say "This is Interesting"". In: *Proceedings of the IEEE Symposium Information Visualization (InfoVis).* IEEE Computer Society, 1996, pp. 54–60. DOI: 10. 1109/INFVIS.1996.559216 (cited on page 150).

[WWK00]   Wang Baldonado, M. Q., Woodruff, A., and Kuchinsky, A. "Guidelines for Using Multiple Views in Information Visualization". In: *Proceedings of the Conference on Advanced Visual Interfaces (AVI).* ACM Press, 2000, pp. 110–119. DOI: 10.1145/345513. 345271 (cited on page 65).

[XW05]   Xu, R. and Wunsch, D. C. "Survey of Clustering Algorithms". In: *IEEE Transactions on Neural Networks* 16.3 (2005), pp. 645–678. DOI: 10.1109/TNN.2005.845141 (cited on page 265).

[Yi+07]   Yi, J. S., ah Kang, Y., Stasko, J. T., and Jacko, J. A. "Toward a Deeper Understanding of the Role of Interaction in Information Visualization". In: *IEEE Transactions on Visualization and Computer Graphics* 13.6 (2007), pp. 1224–1231. DOI: 10.1109/ TVCG.2007.70515 (cited on page 132).

[Yu+12]   Yu, L., Efstathiou, K., Isenberg, P., and Isenberg, T. "Efficient Structure-Aware Selection Techniques for 3D Point Cloud Visualizations with 2DOF Input". In: *IEEE Transactions on Visualization and Computer Graphics* 18.12 (2012), pp. 2245–2254. DOI: 10.1109/TVCG.2012.217 (cited on page 158).

[Zgr+17]   Zgraggen, E., Galakatos, A., Crotty, A., Fekete, J.-D., and Kraska, T. "How Progressive Visualizations Affect Exploratory Analysis". In: *IEEE Transactions on Visualization and Computer Graphics* 23.8 (2017), pp. 1977–1987. DOI: 10.1109/TVCG.2016.2607714 (cited on page 303).

[ZH16]   Zhou, L. and Hansen, C. D. "A Survey of Colormaps in Visualization". In: *IEEE Transactions on Visualization and Computer Graphics* 22.8 (2016), pp. 2051–2069. DOI: 10.1109/TVCG.2015. 2489649 (cited on page 127).

Index

Figure Credits

Figure 1.5 Reprinted by permission from Springer Nature Customer Service Centre GmbH: Radloff, A. et al. "Smart Views in Smart Environments". In: *Proceedings of the Smart Graphics*. Springer, 2011, pp. 1–12. DOI: 10.1007/978-3-642-22571-0_1, © 2011.

Figure 3.24a and Figure 3.28 (central part) Reprinted by permission from Springer Nature Customer Service Centre GmbH: Schumann, H. and Müller, W. *Visualisierung: Grundlagen und Allgemeine Methoden*. Springer, 2000. DOI: 10.1007/978-3-642-57193-0, © 2000.

Figure 3.24b © 2005 IEEE. Reprinted, with permission, from Schumann, H. and Müller, W. *Visualisierung: Grundlagen und Allgemeine Methoden*. Springer, 2000. DOI: 10.1007/978-3-642-57193-0.

Figure 3.26 (main part) © 2007 IEEE. Reprinted, with permission, from Luboschik, M. and Schumann, H. "Explode to Explain – Illustrative Information Visualization". In: *Proceedings of the International Conference Information Visualisation (IV)*. IEEE Computer Society, 2007. DOI: 10.1109/IV.2007.50.

Figure 3.38 © 1999 IEEE. Reprinted, with permission, from van Wijk, J. J. and van Selow, E. R. "Cluster and Calendar Based Visualization of Time Series Data". In: *Proceedings of the IEEE Symposium Information Visualization (InfoVis)*. IEEE Computer Society, 1999, pp. 4–9. DOI: 10.1109/INFVIS.1999.801851.

Figure 4.49 © 2015 Morgan & Claypool. Reprinted with kind permission from Tominski, C. *Interaction for Visualization*. Synthesis Lectures on Visualization 3. Morgan & Claypool, 2015. DOI: 10.2200/S00651ED1V01Y201506VIS003.

Figure 5.1 © 2008 IEEE. Reprinted, with permission, from Bachthaler, S. and Weiskopf, D. "Continuous Scatterplots". In: *IEEE Transactions on Visualization and Computer Graphics* 14.6 (2008), pp. 1428–1435. DOI: 10.1109/TVCG.2008.119.

Figure 5.3 © 2006 IEEE. Reprinted, with permission, from Novotny, M. and Hauser, H. "Outlier-Preserving Focus+Context Visualization in Parallel Coordinates". In: *IEEE Transactions on Visualization and Computer Graphics* 12.5 (2006), pp. 893–900. DOI: 10.1109/TVCG.2006.170.

Figures 5.9 and 5.10 © 2014 IEEE. Reprinted, with permission, from Abello, J. et al. "A Modular Degree-of-Interest Specification for the Visual Analysis of Large Dynamic Networks". In: *IEEE Transactions on Visualization and Computer Graphics* 20.3 (2014), pp. 337–350. DOI: 10.1109/TVCG.2013.109.

Figures 5.43 and 5.44 Reprinted by permission from Springer Nature Customer Service Centre GmbH: Aigner, W. et al. *Visualization of Time-Oriented Data*. Springer, 2011. DOI: 10.1007/978-0-85729-079-3, © 2011.

Figure 6.1 Reprinted by permission from Springer Nature Customer Service Centre GmbH: Eichner, C. et al. "A Novel Infrastructure for Supporting Display Ecologies". In: *Advances in Visual Computing: Proceedings of the International Symposium on Visual Computing (ISVC)*. Springer, 2015, pp. 722–732. DOI: 10.1007/978-3-319-27863-6_68, © 2015.

Figure 6.4 © 2012 IEEE. Reprinted, with permission, from Radloff, A. et al. "Smart Interaction Management: An Interaction Approach for Smart Meeting Rooms". In: *Proceedings of the Eighth International Conference on Intelligent Environments (IE)*. IEEE Computer Society, 2012, pp. 228–235. DOI: 10.1109/IE.2012.34.

Figure 6.12 © 2012 IEEE. Reprinted, with permission, from Streit, M. et al. "Model-Driven Design for the Visual Analysis of Heterogeneous Data". In: *IEEE Transactions on Visualization and Computer Graphics* 18.6 (2012), pp. 998–1010. DOI: 10.1109/TVCG.2011.108.

Figures Licensed Under Creative Commons License

The figures listed in the following are licensed under the Creative Commons Attribution 4.0 International License. To view a copy of this license, visit creativecommons.org/licenses/by/4.0/.

Figures 3.12, 3.51, 5.15, 5.16, 5.19 to 5.21, 5.32 and 5.33 by Martin Röhlig.

Figure 3.25 by Thomas Nocke.

Figures 3.46 and 3.50 by Steve Dübel.

Figure 5.4 by Helwig Hauser.

Figures 5.8, 5.40 and 5.41 by Steffen Hadlak.

Figure 5.12a and Figures 5.13, 5.14, 6.2 and 6.6 by Christian Eichner.

Figures 5.22, 5.28 and 5.29 by Martin Luboschik.

Figure 6.20 by Marco Angelini.

Figure 6.24 by Axel Radloff-Delosea.

Availability of Original Figures

All original figures from this book are released under the Creative Commons Attribution 4.0 International License (CC BY 4.0). The figures are available on `https://ivda-book.de`.